Special Issues in Data Management

ACS SYMPOSIUM SERIES 1110

Special Issues in Data Management

Norah Xiao, Editor
University of Southern California
Los Angeles, California

Leah Rae McEwen, Editor
Cornell University
Ithaca, New York

**Sponsored by the
ACS Division of Chemical Information**

American Chemical Society, Washington, DC

Distributed in print by Oxford University Press, Inc.

Library of Congress Cataloging-in-Publication Data

Special issues in data management / Norah Xiao, editor, University of Southern California, Los Angeles, California, Leah Rae McEwen, editor, Cornell University, Ithaca, New York ; sponsored by the ACS Division of Chemical Information.
    pages cm. -- (ACS symposium series ; 1110)
Includes bibliographical references and index.
ISBN 978-0-8412-2712-5 (alk. paper)
    1. Digital libraries--Management--Great Britain--Congresses. 2. Digital libraries--Management--United States--Congresses. 3. Science and technology libraries--Congresses. 4. Database management--Congresses. 5. Chemistry--Information services--Congresses. I. Xiao, Norah, editor of compilation. II. McEwen, Leah Rae, editor of compilation. III. American Chemical Society. Division of Chemical Information, sponsoring body.
ZA4080.S64 2012
025.040941--dc23
                  2012039247

The paper used in this publication meets the minimum requirements of American National Standard for Information Sciences—Permanence of Paper for Printed Library Materials, ANSI Z39.48n1984.

Copyright © 2012 American Chemical Society

Distributed in print by Oxford University Press, Inc.

All Rights Reserved. Reprographic copying beyond that permitted by Sections 107 or 108 of the U.S. Copyright Act is allowed for internal use only, provided that a per-chapter fee of $40.25 plus $0.75 per page is paid to the Copyright Clearance Center, Inc., 222 Rosewood Drive, Danvers, MA 01923, USA. Republication or reproduction for sale of pages in this book is permitted only under license from ACS. Direct these and other permission requests to ACS Copyright Office, Publications Division, 1155 16th Street, N.W., Washington, DC 20036.

The citation of trade names and/or names of manufacturers in this publication is not to be construed as an endorsement or as approval by ACS of the commercial products or services referenced herein; nor should the mere reference herein to any drawing, specification, chemical process, or other data be regarded as a license or as a conveyance of any right or permission to the holder, reader, or any other person or corporation, to manufacture, reproduce, use, or sell any patented invention or copyrighted work that may in any way be related thereto. Registered names, trademarks, etc., used in this publication, even without specific indication thereof, are not to be considered unprotected by law.

PRINTED IN THE UNITED STATES OF AMERICA

Foreword

The ACS Symposium Series was first published in 1974 to provide a mechanism for publishing symposia quickly in book form. The purpose of the series is to publish timely, comprehensive books developed from the ACS sponsored symposia based on current scientific research. Occasionally, books are developed from symposia sponsored by other organizations when the topic is of keen interest to the chemistry audience.

Before agreeing to publish a book, the proposed table of contents is reviewed for appropriate and comprehensive coverage and for interest to the audience. Some papers may be excluded to better focus the book; others may be added to provide comprehensiveness. When appropriate, overview or introductory chapters are added. Drafts of chapters are peer-reviewed prior to final acceptance or rejection, and manuscripts are prepared in camera-ready format.

As a rule, only original research papers and original review papers are included in the volumes. Verbatim reproductions of previous published papers are not accepted.

ACS Books Department

Contents

Preface ix

1. Diversions and Distractions on the Path to Effective Research Data Curation 1
   Graham Pryor

2. Hosting a Compound Centric Community Resource for Chemistry Data 19
   Antony J. Williams

3. Supplemental Journal Article Materials 31
   David P. Martinsen

4. National Data Management Initiatives and the U.S. Exemplar: DataONE 47
   Suzie Allard

5. Activities of Regional Consortia in Planning e-Science Continuing Education Programs for Librarians in New England 69
   Donna Kafel

6. Interdisciplinary Data Science Education 97
   Jeffrey Stanton, Carole L. Palmer, Catherine Blake, and Suzie Allard

7. Data Management Services in Libraries 115
   Patricia Hswe

8. Research Data Management and the Role of Libraries 129
   Mary C. Schlembach and Carol A. Brach

9. Preparing To Support Research Data Sharing 145
   Ye Li and Lori Tschirhart

Editors' Biographies 163

Indexes

Author Index 167

Subject Index 169

Preface

This book opens discussion on the topics of data management. It is not a new topic at all because scientists deal with data every day; what makes this topic unique is the scale of data everyone is dealing with nowadays and the support that it needs. It is no longer an individual initiative but a common interest and collaborative effort for everyone involved in the life cycle of data, information, and knowledge, who might be researchers, librarians/information specialists, policy makers, or administrators. This book originated from the American Chemical Society (ACS) National Meeting Chemical Information Division (CINF) Symposium entitled "Data Archiving, E-Science and Primary Data" in Anaheim, California on March 28, 2011. It was organized to "explore the challenges and opportunities of supporting e-science research and data management in research libraries" with particular interest in "current applications and practices and preparation opportunities for information professionals". Only one author from the original symposium was able to contribute. With this interesting twist, the book now brings new and refreshing perspectives to the topic by authors from different sectors with diverse background and experience; researchers in science, information science and data management, librarians, and publishers.

Every chapter is independent, with the author's own point of view on the topic; therefore, the reader may begin this book by reading the chapters in any order. Every chapter, like all chapters of the ACS Symposium Series, is carefully peer reviewed and revised before publication. We sincerely hope that our book will be a valuable resource for people interested in the topic. Moreover, we thank each of our authors for their expertise and contributions to this book!

This book would not have been possible without the dedication and effort of all our peer reviewers, my co-editor, Ms. Leah McEwen, and everyone at the ACS Books Department, as well as all of our professional colleagues in the library and information science field. It has been my great pleasure to work with all of you on this project and in the field. My professional life is so interesting, challenging, and fruitful because of all of you! Happy reading!

Norah Xiao

ACS Journal Publishing Group
August 2012

Chapter 1

Diversions and Distractions on the Path to Effective Research Data Curation

Graham Pryor*

Digital Curation Centre, Edinburgh EH8 9LE, United Kingdom
*E-mail: graham.pryor@ed.ac.uk

The question we need to ask before driving everyone down the road to data curation nirvana is why do it at all and, in corollary, what are the consequences from not doing it? Those who will be most engaged in the doing of it, whether researchers, librarians, informaticians or other faculty support, already have to contend with a barrage of competing calls on their time and their budgets, so why should they welcome the additional burden implied by demands that they curate and share their data? This chapter starts by examining the key policy and business drivers for doing just that, identifying the synergies that management strategies may have with the cause of data curation and data sharing, and considering the extent to which they coincide with the increasingly prescriptive policies of the major research funders, principally in the UK. Setting these two perspectives in the context of the traditional research culture, I conclude by drawing some inferences concerning those actions essential to achieving any motivational impetus for adopting new practices in the curation and sharing of data across the global research community.

A Routine Practice

The UK's Digital Curation Centre (DCC) can provide many examples of good practice in research data management from across the realm of higher education, yet we are continuing to apply a significant level of our resource to basic programmes of advocacy. From our rolling regional roadshows (http://www.dcc.ac.uk/events/data-management-roadshows) (*1*) to our series

© 2012 American Chemical Society

of focused institutional engagements (http://www.dcc.ac.uk/community/institutional-engagements) (*2*), each one an intensive sixty day cycle of tailored support aimed at increasing capability in the development of data curation strategies and services, we are still finding it necessary to explain the new forces that are shaping research practice, the advantages of planning ahead in order to meet them and turn them to advantage, as well as the expectation that if compliance with regulation is not met there could be very real consequences.

We need to understand why, despite a recognizable increase in awareness amongst particular cohorts of our research community, such a need for advocacy persists. It is some years since communications and information technology first enabled the digital age and in the heady world of academia new methods and practices in the conduct of research have always found an enthusiastic body of early adopters; so that, by now, shouldn't one have expected the effective organization and management of digital output to have become routine? The DCC was itself founded in March 2004, following a realization by the community's senior stakeholders that the establishment of a national centre for solving challenges in digital curation was essential if individual institutions or disciplines were to be relieved of that already conspicuous burden. Three years later the UK Data Archive (UKDA) (http://www.data-archive.ac.uk/) celebrated forty years as the curator of the largest collection of digital data in the social sciences and humanities in the UK. The UKDA is of course just one of several national data centres serving specific discipline groups, which demonstrates how active data curation was nothing new to particular sectors of the research community.

So, in the context of UK higher education, attention had been drawn formally to the need for dealing with the challenges of research data management, and it was only a matter of months after the DCC had been funded for a second three year term that, in 2008, Research Councils UK, a strategic partnership of the UK's seven research councils, published its Policy and Code of Conduct on the Governance of Good Research Conduct (http://www.rcuk.ac.uk/Publications/researchers/Pages/grc.aspx) (*3*). This document explained, *inter alia*, and for the benefit of all UK research organisations funded by the seven, that unacceptable research conduct includes the mismanagement or inadequate preservation of data. RCUK's agenda for good practice requires institutions to keep clear and accurate records of their research; to hold those records securely in paper or electronic form; to make primary data and research evidence accessible to others for reasonable periods after the completion of any research; to manage data according to the research funder's data policy and with respect to all relevant legislation; and, wherever possible, to deposit data permanently within a national collection.

As explained in the welcome to RCUK's home page, "each year the Research Councils invest around £3 billion ($4.7 billion) in research covering the full spectrum of academic disciplines". One might then have assumed that their Policy and Code of Conduct, with its guidelines for managing research output, would have become the canonical memorandum for all researchers and institutional research managers, the platform upon which good practice should be built. From the reports of numerous surveys of researcher practice conducted over the past decade, patently, it has not. Was the flaw in RCUK's assertion that "responsibility for proper management and preservation of data and primary materials is shared between the researcher and the research organization" (*5*)? Was it simply the absence of a big stick that reduced the impact of these fairly basic measures? One might think so, given the attention claimed by a more recent pronouncement.

Money Talks

In 2011, the Engineering and Physical Sciences Research Council (EPSRC), which at £385 million ($600 million) is the largest public provider of research grants in the UK, published its Policy Framework on Research Data (http://www.epsrc.ac.uk/about/standards/researchdata/Pages/default.aspx) (*4*). It differs from the data policies of the other public funding councils in that the EPSRC does not require researchers to include a data management plan with their grant submissions. Neither, unlike the Natural Environment Research Council (NERC) or the Economic and Social Research Council (ESRC), does the EPSRC provide researchers with a dedicated publications or data repository. Instead, researchers are expected to utilise whatever institutional or subject-based repositories are available to them; where these do not exist, their research institutions are expected to take steps to preserve their data securely.

At first glance it may seem that this is a soft option for the research council. Taking an apparently hands-off, almost laissez-faire approach, responsibility for servicing the complex and expensive undertaking of research data management is openly delegated to those universities in receipt of EPSRC funding. But whilst that responsibility is made explicit, what really makes the EPSRC policy framework different is the requirement that those institutions it funds must have developed "a clear roadmap to align their policies and processes with EPSRC's expectations by 1st May 2012", with a view to being "fully compliant with these expectations by 1st May 2015.

The nine expectations (http://www.epsrc.ac.uk/about/standards/researchdata/Pages/expectations.aspx) (*6*) themselves are not inconsiderable. They charge research organisations with the responsibility to promote internal awareness of the principles upon which the expectations are founded, as well as of the prevailing regulatory environment; they require that published research papers include a short statement describing how and on what terms any supporting research data may be accessed; even publicly-funded research data that is not generated in digital format has to be stored in a manner to facilitate it being shared. Most notably, too, there are specific requirements for the assignment of appropriately structured metadata and digital object identifiers, and for the provision of secure preservation services. These prescriptions will, it is asserted, apply throughout the complete data lifecycle, where the full range of responsibilities associated with data curation over that lifecycle must be clearly allocated within the research organisation.

EPSRC will monitor progress and compliance on a case by case basis. It will investigate non-compliance and, if it appears that proper sharing of research data is being obstructed, EPSRC reserves the right to impose appropriate sanctions. If, after the 2015 deadline, an institution is found to be deliberately obstructing the proper sharing of research data, or otherwise seriously failing to comply with

EPSRC's expectations for managing data, those sanctions could result in the institution being declared ineligible for EPSRC support.

As the largest source of research grant funding, any statement like that coming from the EPSRC is bound to warrant careful consideration. (Figure 1 shows just how much money was being distributed in the 2010-11 academic year and to how many UK institutions.) But is the pursuit of funding, or the risk of losing access to it, the principal and over-riding reason for researchers or research organisations to put effort into curating and sharing research data?

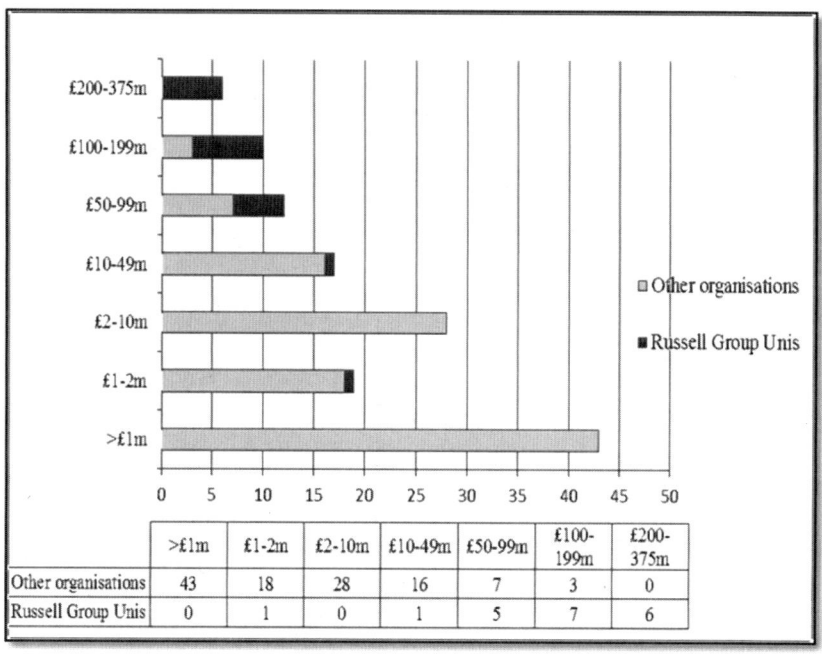

Figure 1. Value of EPSRC research grants on 1 November 2010 by number and type of organization in receipt. (reproduced by permission of the DCC; licensed under a Creative Commons Attribution 2.5 UK: Scotland License)

The Paradox of Competition

Here one also finds a dichotomy, since financial considerations may actually explain why institutions have been slow to tackle the data deluge. The EPSRC's ninth expectation declares that "research organisations will ensure adequate resources are provided to support the curation of publicly-funded research data; these resources will be allocated from within their existing public funding streams, whether received from Research Councils as direct or indirect support for specific projects or from higher education Funding Councils as block grants" (6). This echoes the RCUK's Common Principles on Data Policy (http://www.rcuk.ac.uk/research/Pages/DataPolicy.aspx), which state that "It is appropriate to use public funds to support the management and sharing of publicly-funded research data" (7). The key word in the EPSRC declaration is "existing". There is no new treasure chest to be opened and the only mechanism for negotiating sufficient funding for data curation is to include the full predicted cost within a research proposal. All well and good when this is standard practice; but it is not, despite four members of RCUK, along with the Wellcome Trust, stating categorically that costs associated with data management and sharing can be included in grant proposals (8), and it will be a brave research team that breaks ranks and risks making its proposals uncompetitive by fully incorporating those costs. This is particularly the case when, according to the authoritative Keeping Research Data 2 report (9), the cost of managing data is weighted towards the earlier stages of the research lifecycle, involving data acquisition, ingest and access.

In any case, the story does not end with the conclusion of a fully funded research project. To continue with the example of the EPSRC data policy framework, in expectation number seven it is stated that research data must be securely preserved for a minimum of ten years from the date that any researcher's 'privileged access' period expires or, if others have accessed the data, from the last date on which access to the data was requested by a third party. Whilst it has been observed that the costs of archival storage and preservation of research data "are consistently a very small proportion of the overall costs and significantly lower than the costs of acquisition/ingest or access activities" (9), as any data curator will know, ten years or potentially twice that (or more) when measured in terms of the active management of digital data will still not come cheaply. The cost of the necessary long term infrastructure could not easily be set against the grant for a finite piece of research and would have to be met by the institution. This is especially challenging in a climate of budgetary constraint. Siphoning off funds to provide research data management services is not a simple matter when institutional managers are having to reduce staff numbers and, in some cases, close departments. Does that explain the slow growth of data awareness in these communities? That it simply isn't being pushed as a priority?

Business Aspirations and Incentives

The main driver for the imposition of data management and sharing policies by RCUK member councils, as well as by major non-Governmental funders such as the Wellcome Trust, is almost universally a desire to extract the most value from their investment in research, which may be realised as public benefit or the enrichment of research itself. In the USA, the National Science Foundation (http://www.nsf.gov/bfa/dias/policy/dmp.jsp), with a budget in financial year 2011 of about $6.8 billion for support to around 20% of federally funded research in America's colleges and universities, also appears to be driven by comparable considerations, requiring investigators "to share with other researchers, at no more than incremental cost and within a reasonable time, the primary data…gathered in the course of work under NSF grants" (10).

But what drives the senior management of a university and is it compatible with these aspirations to manage and share research data? Principally, there are two key drivers that most exercise the minds of university management teams: the acquisition of sufficient funding to ensure their institution's sustainability, coupled with the ambition to achieve best in class status for their university or college (or as near to best as it can reasonably aspire) in teaching and, more especially, in its research profile.

Quite uniquely in the UK, both aims are made intrinsically self-supporting by the national structure for evaluating research performance, for it is the Research Excellence Framework (REF) (http://www.ref.ac.uk/) (*11*) that provides a mechanism for assessing the quality of research undertaken in universities and whose assessments are then used to inform the selective allocation of research funding to institutions of higher education. Four panels of assessment between them cover the gamut of research disciplines and submissions to the REF will, as has been the tradition for previous assessment regimes, include details of publications and other forms of assessable output. However, only one of those panels, Main Panel C (http://www.ref.ac.uk/pubs/), which covers a broad range of social sciences, has declared that submissions can include "Digital artefacts such as data sets, multi-use data sets, archives, software, film and other non-print media, web content such as interactive tools" (*12*). For the other three panels, research data has not been explicitly identified as a recognised output type.

This is somewhat bizarre when one of the three primary purposes of the REF is to produce evidence of the benefits of public investment in research. If that is the case, and since the REF is overtly and controversially designed to establish evidence of impact from research, why not a greater focus on the output that has already been recognised by the research councils themselves as having such promise when effectively managed, shared and preserved? Is it a question of definition? Certainly a clear understanding of terminology is an issue that seems to be confounding some institutions engaging with the DCC, where data management steering groups frequently challenge us to explain what actually comprises the research data they should more actively be managing. Is it the raw experimental output from laboratory work, they ask, or is it processed data; and if only the latter needs to be curated, what should be done with the experimental data? Or is it only the data that is eventually distilled and used to support a publication that they need to focus upon? Is that then the problem – that the meaning of what has to be managed and measured is so ill determined, seemingly nebulous, and destined as a consequence for the too difficult folder?

For university management in the UK, before the EPSRC wake-up call, these questions may not have seemed pressing. To establish research data management strategies or, more especially, research data management services for those disciplines not served by national centres, was not only a complex question but one liable to require an answer involving significant investment in both technological and human infrastructures. With budgets shrinking in value any new initiative in this direction was likely to require a politically difficult strategic change in terms of the distribution of internal financing and in structural reorganization. At a time when "Leaders of the UK's most prestigious universities have warned that government plans to cut funding will lead to a higher education meltdown" (*13*), who can blame them for delaying?

The Desire for Impact

But with the first REF looming large and due to be completed in 2014, when impact will count for 20% of an assessment (increasing in subsequent REFs), the scramble to identify measurable impact from amongst the torrent of data generated must surely provide an impetus and incentive for researchers to exercise more forethought when planning their research. The Beyond Impact project (http://beyond-impact.org/), funded by the Open Society Foundation, aimed to "facilitate a conversation between researchers, their funders, and developers about what we mean by the 'impact' of research and how we can make its measurement more reliable, more useful, and more accepted by the research community" (14). Contributors to the project were also concerned that non-traditional outputs from research, often synonymous with data sets, whilst having considerable demonstrable impact were not counting when it came to performance measurement, as has been referred to already.

At a workshop in May 2011, the Beyond Impact project developed the concept of an 'enhanced personal impact dashboard', a container for researchers to list their entire range of research outputs as a live CV, from which an aggregate score could be created, thus enabling a broader measure of impact than is achievable from the consideration of published books, articles and reports. In this somewhat rarified conversation between researchers, funders and developers perhaps there is the seed of a solution to the question of what motivates researchers to manage their data and make it more visible. But is this going to be enough to encourage other than the minority of enthusiasts to take an active interest in data curation?

A series of case studies of researchers in the life sciences, published in late 2009, observed the distance that would have to be travelled to reach effective levels of curation, remarking that in contrast with the disciplines handling 'big data', as in the fields of high energy physics or in genomics, "Data curation is…only one element in the research lifecycle, and did not feature prominently in our case studies" (15). In most cases they simply lacked the motivation and, in any case, competing pressures on their time militated against finding opportunities to address it. They saw themselves as "securing career rewards for the research they do…rather than the data they collect. They are reluctant to share the data that make up their 'intellectual capital'" (16). Further, "The value that possessing particular kinds of information and experience presents to an individual or group is linked to their ability to trade possession to advantage in terms of reputation, funding and career development" (17). That sense of trade as an incentive for researchers to undertake or negotiate services for data curation and sharing is surely somewhere at the core of the, as yet, unattained answer as to why researchers would allow it any of their precious time and effort.

Open to All?

There is, of course, a body of opinion that believes openness to be the heart beating in the breast of successful research. The Panton Principles (http://pantonprinciples.org/), officially launched in February 2010, declared that "For science to effectively function, and for society to reap the full benefits from scientific endeavours, it is crucial that science data be made open" (*18*), where openness is defined in terms of any kind of material or data output from research. The Panton Principles advocate that "data in science should be freely available on the public internet permitting any user to download, copy, analyse, re-process, pass them to software or use them for any other purpose without financial, legal, or technical barriers other than those inseparable from gaining access to the internet itself".

Of course, to achieve such an ideal of openness and the consequent usability of data demands that it must be organised, managed and prepared in such a way that it is (and will remain) discoverable, accessible and reusable. In other words, it will be properly curated. Yet in the report Open to All? (http://www.rin.ac.uk/our-work/data-management-and-curation/open-science-case-studies), a collection of case studies of openness in research commissioned by the Research Information Network and the National Endowment for Science, Technology and the Arts, it was reported that "sustaining and making use of the new kinds of infrastructure required for openness demands new skills and significant effort from researchers and others, particularly at a point when standards, guidelines, conventions and services for managing and curating new kinds of material are as yet under-developed and not always easy to use" (*19*).

At this point, with such a diversity of views and tangential opinion having been uncovered, it is worth refreshing the fundamental question posed at the outset to this chapter: why curate and share research data in the first place and what are the consequences from not doing it? As we have seen, some believe it is worth doing, others are highly committed, although the reasons given are not always closely replicated across individual stakeholder groups.

For a considerable number of years the UK has benefited from having national data centres engaged in the preservation and supply of data from specific discipline groups, including most prominently the social, environmental and archaeological sciences. Their mission is to gather and disseminate the scholarly record represented by authenticated scientific data, as a product or resource having lasting value. Typically, the ESRC, which funds the UKDA, affirms in its research data policy (http://www.esrc.ac.uk/about-esrc/information/data-policy.aspx) that "Data are the main asset of economic and social research. We recognise publicly-funded research data as valuable, long-term resources that, where practical, must be made available for secondary scientific research" (*20*). Recently, others amongst the major research funders have subscribed in a more assertive fashion to that mission; but also, in a climate of increasing public accountability, with an eye on defending their investment of taxpayers' money. Allied to them are the researchers who see good data management coupled with openness and sharing as a means rather than a threat to the development of their research (and hence career) profiles, together with some more altruistic intentions for the betterment of research as a cause in its own right. As for institutional management, their interest has at last been awoken by the increasingly competitive nature of the struggle for research funding in an environment where the overall pot is reducing in real terms whilst the number of institutions wanting access to it have been increasing since the passing of the Further and Higher Education Act of 1992.

Self Defining Data Dependency

As I have already intimated, if only in passing, some disciplines whose business is more obviously datacentric have confidently embarked on their own data management regimes. For example, a typical technical design report prepared for the ATLAS high energy physics collaboration in 2005 provides detailed plans for accessing and processing data on a global basis, covering the approach to be taken and the necessary architecture for both data and data management. This is early data management planning conducted on a large scale. As acknowledged in the report, it is based on experience acquired over many years and is a natural component of project planning where that project is all about data (*21*). Yet it is admitted by those working in high energy physics research that whilst "Data from high-energy physics (HEP) experiments are collected with significant financial and human effort and are in many cases unique...At the same time, HEP has no coherent strategy for data preservation and reuse, and many important and complex data sets are simply lost" (*22*). Hence, a study group on data preservation and long-term analysis (DPHEP) has been formed and a series of workshops held to investigate this issue in a systematic way.

This is a large cohort of researchers owning full recognition of the actions needed and with no misperceptions about why they are curating and sharing their data. Nonetheless, such an awareness of good data management as a core and necessary practice in research is limited to particular such groups that are themselves defined by the scope of their data dependency.

Notwithstanding such examples of assuredness, it remains a fact that the DCC was charged in 2010 with the specific mission to go amongst the UK's higher education institutions and help them to build capacity, capability and skills in data management and curation, in its third phase (from March 2010 until February 2013) taking the strategic decision not to attempt to proselytize the huge and diverse community of researchers but instead to focus initially upon the support teams, the professionals who are the data scientists, managers and custodians, the informaticians and data librarians through whom knowledge and techniques may be transferred to the research practitioner (*23*).

Traditions and Techniques

Academia is a landscape of principles, guidelines, initiatives and conversations. Unlike the private sector, it is rarely shaped by direct instruction or command, which would be anathema to the traditions of academic freedom and intellectual rigour. Consequently, it can be slow to change course. One can see fine examples of this cultural difference in the markedly dissimilar reception given first to RCUK's Common Principles on Data Policy, which although lucidly stated and unarguably helpful as "an overarching framework" (*7*) are presented passively, rendering them likely to be relegated to background noise; and then to the direct imperative contained in the EPSRC's "clear expectations of organisations in receipt of EPSRC research funding" (*6*), which have galvanised institutions of all complexions into a fever pitch of activity as they strive to produce their data management roadmaps by the May 2012 deadline.

But what is the imperative that will drive our support teams, the growing band of data professionals? Certainly, as noted by Corrall (*24*), the "management of the research data generated by e-science and e-research has replaced open access to scholarly publications as the hot topic on the academic library and information services agenda". She refers to the earlier observation by Hey and Hey (*25*) that by responding effectively to the challenge "the e-Science revolution will put libraries and repositories centre stage in the development of the next generation research infrastructure".

For librarians in particular, this opportunity to strategically reposition themselves should be welcome. Whilst for centuries they had occupied centre stage as the recognised custodians of documented knowledge, with many 'scholar librarians' acknowledged as members of the research community in their own right, in more recent times there is evidence of a serious disconnection of the researcher from the university library. The report Patterns of Information Use, referred to above, identified a situation in which, empowered by the Internet, "researchers have removed themselves from the mainstream library user population. They do not even use the library catalogue" (*26*). Whilst this report covered the life sciences, subsequent studies in other disciplines have tended to corroborate a view that "the traditional role of professional information intermediaries has been largely replaced by direct access to online resources" (*26*). These findings speak of straitened times for a library profession relegated to supporting the undergraduate body.

Digital curation involves the active management of research data sets, which activity includes a number of tools and skills long familiar to the library community: selection, appraisal, the assignment of metadata and the classification and organisation of knowledge, these traditional library activities together with secure storage are all appropriate to the creation, storing, accessing and rendering of data. The way in which institutional repositories have emerged as a new function of the central information service underwrites how the organisation of digital content is being treated as a collection management issue, albeit these repositories as yet have tended to focus on electronic documents rather than data storage. Yet, despite the obvious synergy, the academic library community in the UK has on the whole been slow to rise to the challenge.

Not so in the USA, where librarians have been quick to re-engineer themselves into a new and highly marketable role. The University of Virginia library (http://www2.lib.virginia.edu/brown/data/), for example, has created a Scientific Data Consulting Group, consisting primarily of repurposed library staff who are working to a new compass point. Whilst the shift in emphasis is potentially significant, they make it appear simple and natural, explaining that "Libraries have been managing information for 4,000 years. Today, your libraries are evolving and building expertise to continue this tradition so that they can help you preserve research data of the past, present, and future" (*27*). The range of services provided is based on a direct response to needs; their benefits to members of the research community are also listed on a tantalising billboard. For the avoidance of doubt they even explain what they mean when they refer to this thing called data:

"Data can mean many different things, but there are typically four main categories that it can be sorted into for management purposes: **Observational**, **Experimental**, **Simulation**, and **Derived or Compiled**. The category that you choose will then have an effect upon the choices that you make throughout the rest of your data management plan" (*28*).

In a situation where, according to the aforementioned Open to All? report, "standards, guidelines, conventions and services for managing and curating new kinds of material are as yet under-developed and not always easy to use" (*19*), librarians equipped to identify and deliver such tools and knowledge do indeed occupy a privileged position, one in which the answer to the question 'why manage data' is for them essentially a matter of professional commitment and, more importantly, sustainability.

Virginia's library is not unique in reaching this conclusion nor in its grasp of the fact that the attention of researchers will only be secured if it can be demonstrated what is in it for them. The University of Maryland libraries, for example, adopt a perspective of informed authority by running workshops for faculty, students, and staff in how to manage experimental data (*29*). There are many more examples of direct action having spread across this professional strata and it has become something of a cause. In December 2012 the DataRes Symposium in Washington DC will be addressing all the broad issues to be met by the academic library community, from how libraries, library and information schools, and data repositories interact with researchers to the challenges in terms of technology and infrastructure that libraries face to support research data management. (*30*). It is a hot topic indeed.

Incrementally the Best of intentions

How that heat is reaching the researcher is a curious matter. The position of their employing institutions, at least in the UK, is less ambivalent and can be shown to have been generally reactive, inspired occasionally by controversies sparked from uncomfortable Freedom of Information requests for access to jealously guarded research data, even incidents of hacking or leaks, the University of East Anglia's 'Climategate' being the first such to achieve notoriety. But in the main this has been a case of dealing with issues of data security rather than data curation, or with policy requirements aimed at enhancing research governance or social and political demands for greater transparency in research. More proactively, the research funders have, both severally and jointly, built new research data policies on the principle that publicly-funded research data are a public good, and therefore demand long-term preservation and high quality data management. In the immediately preceding paragraphs we have also reflected on the attitude of the support groups to the requirement for good data management, most notably that of the information professionals, which has been shown as both responsible and enterprising.

But of all the stakeholders engaged in the business of producing, using and curating research data, we find ourselves returning repeatedly to the question 'what of the researchers themselves?' – and here I am referring to those researchers not already served by the infrastructures of big science, nor those enjoying the support of national data centres. What of this larger body, this enigma, whose apparent lack of awareness of simple good practice in data management can be somewhat surprising?

The Incremental project (http://www.lib.cam.ac.uk/preservation/incremental/), a collaboration between the Universities of Cambridge and Glasgow, in 2010 sought to investigate and improve their institutions' research data management infrastructure. The investigation reported inconsistent structures for creating and organising data, the storage of data on cheap and flimsy media, a failure to use networked back-up services (often resulting in moderate to catastrophic data loss), uncertainty about formats and media for data preservation and certainly no evidence of sufficient time being allowed for the considered assignment of metadata. These are all such basic failures in what should be common practice. When it came to asking about their attitudes to data sharing, while many researchers were positive in principle, they were almost universally reluctant in practice. One even remarked "it's hard to overcome your personal investment; it's like giving away your baby" *(31)*.

Incremental's recommendations very much tended towards the provision of easily digested advice and training, laying the building blocks for a more comprehensive data management infrastructure. Following the investigation phase clear online support has been created and publicised, with a series of workshops taking place at which the benefits of good research data management was explained to their researcher populations. One might argue that the Incremental team's findings and the nature of their consequent restorative actions do much to clarify why the DCC is driven to continue with its programme of advocacy, since one of the project's key messages seems to have been that those responsible within institutions for delivering facilities in support of research data management simply have not done a good job in explaining what is necessary, what should be done and what help is available. Yet, even whilst knowing how poor some data management practices have been, it is also tempting to ask what really are those benefits to researchers that, until now, they seem to have done without quite happily, if oblivious of risk?

Citation: Rewards and Barriers

One serious temptation for researchers would be an opportunity to increase their research profile, with an enhanced level of citation being the natural outcome of effective data management planning, curation and consequent sharing. Or so it is said in numerous claims for the open access publication of scholarly articles, a claim that one might expect to extend to research data. But the available evidence to support this conjecture is limited and the increase in data citations that are being reported are more modest than may have been anticipated. As noted by Piwowar, "Rewarding investigators who share data, assessing the impact of data repositories, and measuring the intended and unintended effects of data policy decisions all depend on being able to track dataset reuse. Unfortunately, tracking data reuse is currently extraordinarily difficult due to diverse attribution practices, tool limitations, and data source restrictions" (*32*).

The problems facing text mining within scholarly articles may also discourage the flight to the practice of openness for data. JISC's report on the Value and Benefits of Text Mining explains that "within academic research, mining and analytics of large datasets are delivering efficiencies and new knowledge in areas as diverse as biological science, particle physics and media and communications"; at the same time, however, current copyright restrictions determine "that the availability of material for text mining is limited" (*33*). If legal uncertainty, inaccessible information silos, an absence of sufficient information and the lack of a critical mass are proving to be barriers to the text mining of articles, as this report would have us believe, there is surely little impetus for preparing and making data available.

Nonetheless, if achievable, the preservation and enhanced discoverability of research data should still be of benefit by enabling the acceleration of dialogues between researchers; it may even open up opportunities for new collaborations and, looking at it from the perspective of research quality, it should also trigger improvements in the standard of published research if the data used to substantiate declarations of new knowledge are made available for close scrutiny, reproduction and testing by informed but independent scholars. Yet it remains to be said that for researchers having to introduce new and effective processes for the management of their data, such a prospect can be viewed as heralding a host of extra burdens that, within the often short horizon of a funded project will threaten to reduce their research capacity.

Such a perception is of course fundamentally erroneous and the consequences of not planning and managing research data are, in any case, of considerably greater magnitude than the inconveniences of avoiding them. Moreover, the increasing sophistication of a ubiquitous, transnational communications and information technology infrastructure has provided an environment where research has become global and data intensive. That is the world of e-research, a given from which we are unlikely to return.

In Making the Case for Research Data Management (http://www.dcc.ac.uk/resources/briefing-papers/making-case-rdm#Drivers) (*34*) Whyte and Tedds explain how funders expect research to be international in scope, with the Royal Society having reported that over a third of all articles published in international journals are internationally collaborative, up from a quarter fifteen years ago (*35*). In such an environment it is a given that researchers will need data management tools and services to function at all. As pointed out in Making the Case,

> "Research data is itself often seen as a form of infrastructure, as it is the basis for 'data intensive' research; a trend spreading from fields such as genomics and astronomy across many domains. As the European Commission Riding the Wave report points out, this trend calls for 'collaborative research data frameworks' (*36*). These should help develop the emerging pan-European collaborative research data infrastructure, and avoid isolating the islands of good practice".

Failing to keep pace with the global research community would be regarded as damaging by all concerned, since both careers and institutions are measured by their international standing. One could add nations to this mix. With an eye to competing strategies elsewhere in the world, the UK Parliament Commons Select Committee for Science and Technology concluded in 2011 that "in order to allow others to repeat and build on experiments, researchers should aim for the gold standard of making their data fully disclosed and made publicly available" (37).

Mixed Motives

There are very evident top-down pressures for seeking to establish good practice in data curation. There are too some particularly strong initiatives from the research coalface, as discussion here of the open science movement and the reorientation of the library professional has indicated. The prospect of a transition to a fully functional research data infrastructure within higher education must now seem ineluctable; it is effectively all part of the struggle for survival. Why it is taking so long to gain momentum in the UK is perhaps a reflection of culture rather than a denial of the inevitable. This could explain the difference between the attitude of academic librarians in the UK compared with those in the USA, where there are no public universities at the national level outside of the military service academies and the influence of an enterprise culture might be expected to be stronger. Hence, perhaps, the speed with which the library community in the USA has responded to the so-called data deluge.

But ignoring all of these forces, whether top down or bottom up, are not the reasons for pursuing the effective curation of research data self-evident to all? There are three key perspectives to be considered when observing the data landscape:

1. *Scale and complexity*, which describes the volume and pace at which research data is being generated (as with the experience of the single sequencer, who can now generate in a day what it took ten years to collect for the Human Genome Project); the technological and human infrastructures necessary to generate, collect, analyse and transport the data globally; and the cultural and organisational dynamics inherent in movements such as that which is pressing for more open science. Bringing coherence and a semblance of order to this situation will require the deployment of standard curation techniques designed first of all to rein in and reduce the problem, starting with the implementation of robust data management plans and the introduction of processes for the selection and appraisal as well as the disposal of non-essential data.
2. *Policy*, an essentially four-cornered reflecting pool containing the sometimes competing demands of government legislation, the aspirations of funders, institutional direction and the often thorny topic of ethics and personal disclosure. Having workable institutional data policies that provide for any predictable contingency, a shared understanding of all the issues by all the stakeholders, together with a support infrastructure

that will inform the community and ensure the continuity and reliability of approach, is a no-brainer in this instance.
3. ***Management***, meaning the active handling of such issues as the choosing of data storage options; the selection, appraisal and disposal of data; dealing with the usually significant costs implied by a data management service; as well of course as the very human issue of incentivising the research community to play ball in what is still to many a game where the risks are legion and the rewards uncertain. Well management is management, and the success of the strategies and decisions taken will be in direct relation to their quality of design and execution. As has already been written, money is the perennially leading topic for management and experts from the world of digital curation can point to cost benefit modellers that will make the issues less opaque.

Meeting the imperatives of perspectives 1 and 2 requires a range of rational responses that are within the grasp of any competent organisation. The financial and technological challenges for management too will find their solutions. But it will be the human issues that will take time; not just the advocacy that is required, the persuasion to do things differently and to understand the advantages to be had from new behaviours. That is ongoing. Since it began its programme of regional data roadshows (http://www.dcc.ac.uk/events/data-management-roadshows) the DCC has been encouraged by the increasing participation of research active staff and senior university managers, which augurs well for the future. Where once we invited ourselves to a location, after less than eighteen months we are receiving requests to stage an event.

But there is one hurdle that remains: how does one incentivise researchers to let go of assured and long-standing methods for attributing recognition and delivering real rewards for their work, only to adopt practices that make new, sometimes substantial demands on their time and with no guarantee that they will have a positive impact on their research and their careers? As we have established, for the career researcher data represents intellectual capital. Why would any of them part with it without receiving something in exchange?

Shortly after I first moved from a management position in the private sector to take up a similar role in academia, which involved negotiating services and policy with academic staff, a colleague asked how I was getting on with herding cats. It was at the time an unfamiliar phrase that, unfortunately, quickly gained a lot of meaningful resonance. Yet there is a solution to the problem of herding cats; you simply leave a bowl of fish at the place where you want them to go. If we are to cut a straight path to effective data curation that bowl will have to contain a clear and indisputable resolution of the recognition and reward conundrum. The rest is almost history.

References

1. Digital Curation Centre. www.dcc.ac.uk/events/data-management-roadshows (accessed 14 May 2012).
2. Digital Curation Centre. www.dcc.ac.uk/community/institutional-engagements. (accessed 14 May 2012).
3. Research Councils UK. www.rcuk.ac.uk/Publications/researchers/Pages/grc.aspx (accessed 14 May 2012).
4. Engineering and Physical Sciences Research Council. www.epsrc.ac.uk/about/standards/researchdata/Pages/default.aspx (accessed 14 May 2012).
5. Research Councils UK. www.rcuk.ac.uk/documents/reviews/grc/goodresearchconductcode.pdf, p8 (accessed 14 May 2012).
6. Engineering and Physical Sciences Research Council. www.epsrc.ac.uk/about/standards/researchdata/Pages/expectations.aspx (accessed 14 May 2012).
7. Research Councils UK. www.rcuk.ac.uk/research/Pages/DataPolicy.aspx (accessed 14 May 2012).
8. Digital Curation Centre. www.dcc.ac.uk/resources/policy-and-legal/overview-funders-data-policies (accessed 14 May 2012).
9. Beagrie, N.; Lavoie, B.; Woollard, M. *Keeping Research Data Safe 2*; JISC, 2010.
10. National Science Foundation. www.nsf.gov/bfa/dias/policy/dmp.jsp (accessed 14 May 2012).
11. REF 2014. www.hefce.ac.uk/research/ref/ (accessed 14 May 2012).
12. REF 2014. www.hefce.ac.uk/research/ref/pubs/2012/01_12/01_12_2C.pdf (accessed 14 May 2012).
13. BBC News Channel. news.bbc.co.uk/1/hi/education/8453360.stm (accessed 14 May 2012).
14. Beyond Impact. beyond-impact.org/ (accessed 14 May 2012).
15. *Patterns of Information Use and Exchange*, a report by the Research Information Network and the British Library; 2009, p 49.
16. *Patterns of Information Use and Exchange*, a report by the Research Information Network and the British Library; 2009, p 38.
17. *Patterns of Information Use and Exchange*, a report by the Research Information Network and the British Library; 2009, p 41.
18. Panton Principles. pantonprinciples.org/ (accessed 14 May 2012).
19. Open To All? Case Studies of Openness in Research, a joint RIN/NESTA report; 2010, p 4.
20. Economic and Social Research Council. www.esrc.ac.uk/about-esrc/information/data-policy.aspx (accessed 14 May 2012).
21. Atlas Computing Technical Design Report, ATLAS TDR-017; June 2005.
22. South, D. M. *Data Preservation in High Energy Physics*; CHEP, 2010.
23. Digital Curation Centre. www.dcc.ac.uk/about-us/dcc-phase-3 (accessed 14 May 2012).
24. Corrall, S. In *Managing Research Data*; Pryor, G., Ed.; Facet Publishing, 2012, p 105.

25. Hey, T.; Hey, J. E-science and its implications for the library community. *Library Hi Tech* **2006**, *24* (4), 526.
26. *Patterns of Information Use and Exchange*, a report by the Research Information Network and the British Library; 2009, p 47.
27. University of Virginia Library. www2.lib.virginia.edu/brown/data/ (accessed 14 May 2012).
28. University of Virginia Library. www2.lib.virginia.edu/brown/data/whymanage.html (accessed 14 May 2012).
29. University of Maryland Libraries, www.lib.umd.edu/UES/sw.html, accessed 14th May 2012
30. University of North Texas Libraries. research.library.unt.edu/datares/wiki/Datares_symposium (accessed 14 May 2012).
31. Freiman, L. *Incremental Scoping Study and Implementation Plan*, July 2010. http://www.lib.cam.ac.uk/preservation/incremental/Incremental_Scoping_Report_062010.pdf (accessed 14 May 2012).
32. Piwowar, H. *Tracking Data Reuse: Motivations, Methods, and Obstacles*. www.iassistdata.org/conferences/2011/presentation/2912 (accessed 14 May 2012).
33. *The Value and Benefits of Text Mining*, a Digital Infrastructure Directions report by JISC; 2012, p 3.
34. Whyte, A.; Tedds, J. *Making the Case for Research Data Management*; DCC, 2011, p 2.
35. *Knowledge, Networks and Nations: Global Scientific Collaboration in the 21st Century*; RS Policy Document 03/11; The Royal Society, 2011.
36. Wood, J. *Riding the Wave: How Europe Can Gain from the Rising Tide of Scientific Data*; EC, 2010.
37. United Kingdom Parliament. www.parliament.uk/business/committees/committees-a-z/commons-select/science-and-technology-committee/news/110728-peer-review-published/ (accessed 14 May 2012).

Chapter 2

Hosting a Compound Centric Community Resource for Chemistry Data

Antony J. Williams[*]

Royal Society of Chemistry, Department of Cheminformatics,
904 Tamaras Circle, Wake Forest, North Carolina 27587
[*]E-mail: williamsa@rsc.org

ChemSpider is one of the chemistry community's primary online resources and distinct from many of the other online offerings in that it allows users to participate in expanding, annotatin and curating the data. It is a free resource developed with the primary intention of aggregating and linking chemical structure based information and data across the web. Expanding in content daily it contains over 27 million unique chemical entities and is linked out to well over 400 data sources. ChemSpider allows text and structure-based searches to resource information such as chemical vendors, properties, analytical data, patents, publications and a mass of other related information. It is also the foundation of a series of other related projects for the management of community deposited chemical syntheses, as the basis of an education platform for students and as the host for spectral data serving a spectroscopy teaching resource and game. ChemSpider is an ideal environment for chemists to expose their scientific activities and assist in creating a freely accessible resource for chemistry related information.

© 2012 American Chemical Society

Introduction

"Just Google it. It's on Wikipedia. The information is definitely out there...just search for it." These comments, and other paraphrased forms, are all too common in our everyday lives as the internet has continued to permeate our *de facto* approach to sourcing data and information. It certainly applies to chemistry as methods of finding information about chemicals, chemical processes, chemists and many other related sciences are distributed across the internet. Chemists enter the name of a chemical of interest into a search engine and filter through the hit list hoping to find a result matching their query. Online resources and web-based searches are the dominant approach by which most of us find information nowadays.

There are a myriad of resources (*1*, *2*) online regarding chemical related data and these are provided by chemical vendors, publishers, government databases, grant-funded academic databases, commercial systems and many other hosts. Access to data is, in general, no longer an issue as so much data is now available online. However, the quality of these data, especially in the wilds of the internet, has been questioned (*3*, *4*). In this regard the opportunity exists for the chemistry community to participate in improving and expanding the data available online and with blogs, wikis, the efforts of Open Notebook Science and certain public domain databases (*vide infra*), the contributions are improving the quality and quantity of accessible information.

In recent years a number of chemical compound databases have come online containing anywhere from a few hundred to many millions of chemical structures with associated information. These can be specialized databases or simply information aggregators where data are aggregated *en masse*. These aggregators generally contain tens of thousands to millions of compounds. This article will discuss one of these databases, ChemSpider (*5*, *6*), and our efforts to build a high quality resource of chemical compounds, syntheses and related data linked out across the internet.

ChemSpider: A Structure Centric Hub for the Internet

ChemSpider (*7*, *8*) was developed as a hobby project by a small team and delivered as a free offering to the chemistry community. The online database was released to the public in March 2007 with the declared intention of creating "a structure centric community for chemists". It has since grown into a resource hosting over 27 million unique chemical structures linking together over 400 original data sources. The data were aggregated from various contributors including chemical vendors, commercial database vendors, government databases, publishers, Open Notebook Science participants as well as a number of individual scientists. The database can be searched using alphanumeric text searching of both intrinsic properties (such as molecular formula and molecular weight), as well as predicted, molecular properties and structure/substructure searching. The diversity of searches has expanded since inception to support various types of users such as mass spectrometrists (9) and medicinal chemists. The search system is flexible, fast and provides access to a lot of data integrated to a particular chemical from across the various depositors. A screenshot of the interface and *partial* results obtained for a search on domoic acid is shown in Figure 1.

Figure 1. A partial screen capture of the ChemSpider record following a search for "Domoic Acid" (http://www.chemspider.com/4445428). The chemical record shows the structure, multiple identifiers, an experimental property (solubility), links to multiple articles and additional infoboxes. The entire record spans multiple pages and links to many external data sources and informational websites.

As a result of aggregating data from a multitude of data sources we observed that the sourced data were of various levels of quality and it became obvious rather early in the project that data curation and validation would be necessary. The ChemSpider development team was small and the curation of millions of chemical compound records was recognized as being essentially impossible without asking the data source providers to cleanse their own data or asking the community to provide crowdsourced support in terms of data validation. The ChemSpider interface was enhanced to allow real time curation of the data and, in parallel, a rules-based curation scheme at deposition was introduced to check for chemistry issues such as hypervalency. As a result hundreds of thousands of incorrect identifiers associated with the chemical compounds were removed. The resulting name-structure validated dictionary has been used to provide high precision chemical name entity extraction (*10*) and to support semantic markup of published chemistry articles (*11*). To clarify the size of the "crowd" it should be clarified that it is actually a rather small group and, at the time of writing, less than 200 people ever have either deposited or curated data on ChemSpider. In 2011 only 16025 curations were performed by 116 people. Clearly the level of crowdsourced contribution is rather small but it should be noted that the *quality* of the contributions, as rated by secondary checks to their work, delivers excellent additions to the database.

The curated chemical identifiers (systematic names, trivial names, registry numbers etc) making up the name-structure dictionary are the basis of integrating to a number of services provided by Pubmed (*12*), Google Books and Google Patents. For example, validated chemical names are used to search against the Pubmed database searching *only* against the title and the abstract. In this way a search on Xanax, for example, would retrieve only those articles with Xanax in the title and abstract rather than the many thousands of articles likely mentioning Xanax in the body of the article. In a similar way chemical names are passed to the application programming interfaces (APIs) for both Google Patents and Google Books content structure searchable via ChemSpider (see Figure 2 for an example). As a result a chemist can draw a structure on ChemSpider, the validated chemical names associated with the structure are then used as the search queries against the appropriate APIs to retrieve books, articles and patents in just a couple of seconds...all for free.

Participation in community-based curation encouraged us to add further capabilities for the annotation and expansion of the data. The ability to add analytical data (specifically spectral data and crystallographic information files (CIFs)) to chemical structure records was provided. Users were also given the ability to deposit single chemicals or files containing multiple chemical structures. Over 4000 spectra have been contributed by members of the community and additional data are now added almost daily. The spectral data types include infrared, Raman, mass spectrometric and NMR spectra but the majority are ^{1}H and ^{13}C spectra. The 1D NMR spectra are used as the source data for a spectral game used to teach of NMR spectral interpretation (*vide infra*).

As a result of the success of ChemSpider and because of the aligned common vision of "Advancing the Chemical Sciences", the Royal Society of Chemistry (RSC) acquired ChemSpider in April 2009 (*13*). The original vision of providing a community portal for chemists to source data and information remains in place. After joining RSC ChemSpider was able to integrate RSC content into the database on an ongoing basis by integrating the article production services and updating of ChemSpider into the workflows.

Figure 2. One of multiple pages of patents retrieved from Google Patents utilizing the series of validated names associated with Domoic Acid as the basis of a text search against the Google Patents programming interface.

ChemSpider SyntheticPages

ChemSpider deals with the integration of data and information associated with chemical compounds. While there are numerous commercial reaction databases, there is no free database of chemical syntheses that the community can contribute to or comment on. In order to extend the coverage to syntheses a new database was established known as ChemSpider SyntheticPages (*14*) (CSSP). Seeded with content from the original SyntheticPages website the community is fully responsible for populating the database with their contributions as CSSP is essentially a publishing platform. The system can host multimedia content, spectral data and links to the ChemSpider database. Most importantly data can be deposited by members of the community. Figure 3 shows an example reaction from CSSP.

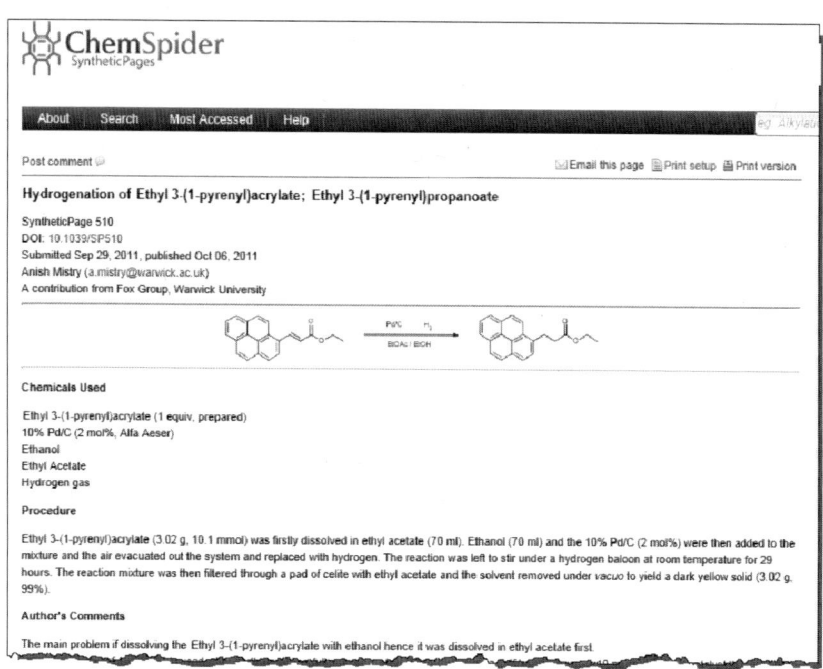

Figure 3. A screenshot showing an example reaction from ChemSpider SyntheticPages.

CSSP provides to chemistry students specifically an ability to develop an online reputation especially since each SyntheticPage has a single author, the chemist who performed the synthesis. After submission a SyntheticPage is reviewed by one or more members of the editorial board and the comments are available to the author, who is informed via email. The author then makes edits online and, when accepted by the editorial board, the article is published. This differs from the classical review process as the original review and feedback is only between the editorial board and the author rather than via an anonymous and extended review process via a set of selected reviewers. As a result the process is generally very fast relative to classical review, commonly less that 48 hours. Until the article is published all exchanges between the editorial board and the author are not visible to the community.

Once published the article can then be commented on by the community. Such comments could be their experiences of repeating the synthesis or alternatives to the synthesis as reported. Each SyntheticPage receives a digital object identifier (DOI) which is a valuable addition to a resume, especially for a student. CSSP hosts hundreds of synthetic procedures and new submissions are made regularly.

Comparing ChemSpider to the Chemical Abstracts Service

Chemical Abstracts Service (*15*) (CAS) is regarded by many as the gold standard in chemistry databases. CAS has been aggregating chemistry-related data for over a century and produces the CAS registry that presently contains over 66 million organic and inorganic substances. In comparison ChemSpider has aggregated over 27 million unique chemical entities with new additions made to the database daily. ChemSpider now inherits new chemicals associated with scientific articles almost daily and before CAS has extracted the data from the publications.

What distinguishes ChemSpider from CAS specifically is that ChemSpider does not abstract data from publications or patents using a team of abstractors. As a result ChemSpider is not growing at the rate that the CAS registry expands. However, this distinction is important since ChemSpider inherits data from non-published sources including depositions from individual scientists and chemistry teams around the world. CAS does not support community-based depositions and has no ability to allow annotation or curation by individuals (even when errors are noted). The integration of ChemSpider to some of the internet-based services such as Google Patents and Books as well as Pubmed, while not equivalent to the extraction and curation of data by CAS, does equate to the delivery of similar types of information. It should be noted that CAS can only be accessed through a paywall while ChemSpider is, of course, totally free access.

ChemSpider for Teaching, Learning, and Research

One of the primary objectives for the RSC is to advance the chemical sciences. This is not only in terms of researchers but also to provide tools with the intention of training the next generation of chemists. To support this mission RSC developed the Learn Chemistry (*16*) platform to provide a central access point and search facility to access various chemistry resources. ChemSpider contains a lot of information of value for students learning Chemistry but also a lot of information not relevant to their studies. The Learn Chemistry wiki (see Figure 4) is a mediawiki environment integrated to ChemSpider and providing access to data and information that is delivered at a level most appropriate to students in their last years of school, and first years of university (ages 16-19). The wiki restricts the list of compounds shown, the properties listed and the spectra and links displayed to those most relevant to studies for this age group. The resource is an interactive resource and allows students to answer a variety of quiz questions, and allowing chemical educators to contribute to the content.

In parallel to the Learn Chemistry wiki the ChemSpider team also manages the development of the SpectraSchool (*17*) website (see Figure 5), a website to learn about various forms of spectroscopy, specifically 1D NMR, mass spectrometry, infrared and UV-Vis spectroscopy. Users have the opportunity to display the various forms of spectral data associated with a number of common organic molecules but also provides a quiz-based mode where the user is shown a number of spectra of various types from which the user has to determine what the chemical compound is.

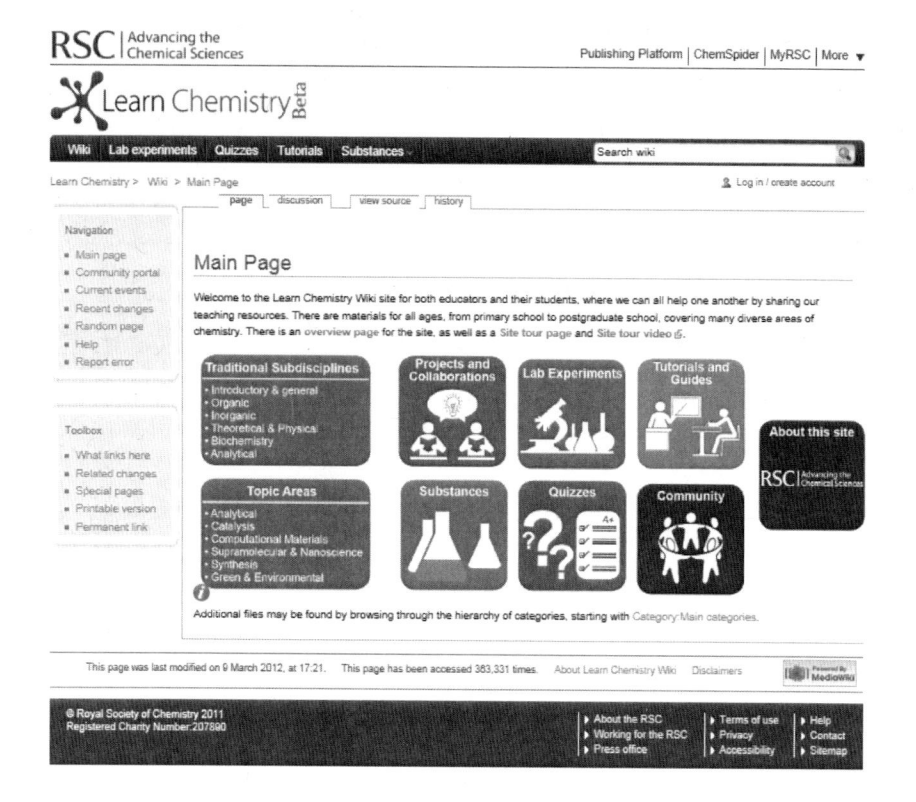

Figure 4. The Learn Chemistry wiki.

The spectroscopic data contained within ChemSpider are also used as the basis for the Spectral Game (*18*), which has already been accessed by over 10,000 students in over 100 countries. The game has students interpret NMR spectra and validate either H1 or C13 spectra against multiple structures. The game increases in complexity as it progresses starting with two structures to choose from and increasing up to a maximum of five chemical structures to choose from. This gaming approach has also helped to identify errors in the spectra, offering an opportunity to data curation through gaming.

ChemSpider SyntheticPages, as discussed earlier, provides an environment for students to populate the online database with their chemical reactions thereby providing the opportunity for students to learn how to write up their procedures, develop their skills for documentation of science and, in parallel develop their reputation in the developing social network for chemistry.

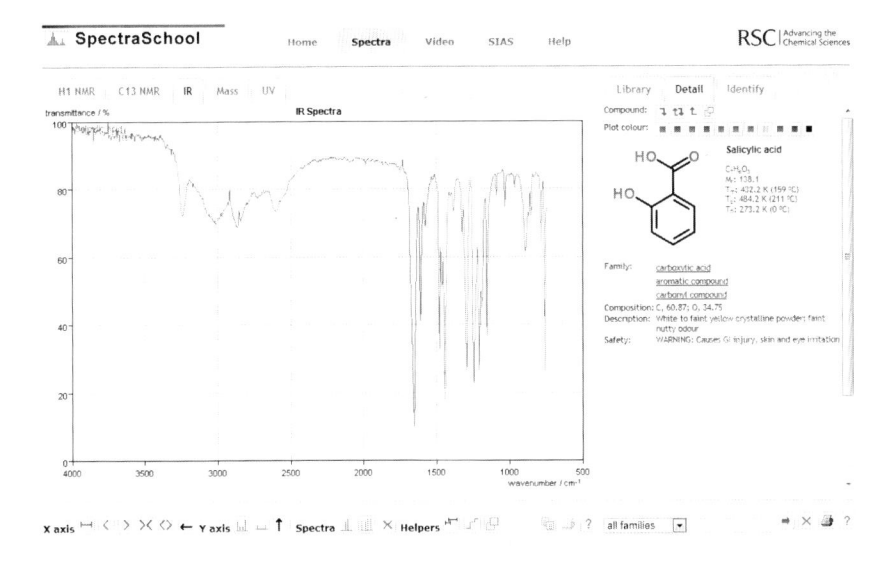

Figure 5. The SpectraSchool website. The infrared spectrum for salicylic acid is displayed. The H1 NMR, C13NMR, IR, MS and UV spectra for salicylic acid can be displayed.

Conclusion

The ChemSpider database has already established itself as one of the premier chemistry sites on the internet and has engaged the community to participate as data depositors and curators. As a resource for students of chemistry the website offers many advantages in terms of sourcing data and accessing information which previously would have been inaccessible except via a library terminal. ChemSpider is the foundation technology under a number of other projects of value to the community, each benefiting from the participation of the community in adding more data. These include the Learn Chemistry wiki and ChemSpider SyntheticPages.

Online chemistry databases such as ChemSpider will continue to become more prominent as the internet provides access to an increasing quantity and diversity of data. Such databases will become increasingly important in supporting both education as well as decision-making processes for researchers by providing access to key data. This increase in freely-accessible data and information will ideally be accompanied by approaches allowing data curation and validation to encourage community contribution to enhancing data quality. In addition, as data-mining tools improve the chemistry databases available online are likely to offer significant opportunities to benefit the discovery process. We live in exciting times where the available technologies encourage collaboration and contribution.

About the Author

Antony Williams is the VP of Strategic Development and Head of the Cheminformatics team at the Royal Society of Chemistry. He has spent over a decade in the commercial scientific software business as Chief Science Officer for Advanced Chemistry Development. He was trained as an NMR spectroscopist by training and has authored over 130 peer-reviewed publications and multiple review articles and book chapters. He continues to focus his passion for providing access to chemistry-related information to the masses with the RSC-ChemSpider team and innovate novel approaches for improved access to chemical data online. He can be found as the ChemConnector on the social networks.

Acknowledgments

ChemSpider results from the work of many people, not only those within RSC. ChemSpider and its associated projects are developed by the cheminformatics team led by Valery Tkachenko (Chief Technology Officer) and includes Colin Batchelor, Aileen Day, Ken Karapetyan, Alexey Pshenichnov, David Sharpe and Jon Steele. All projects are supported by a dedicated team of IT specialists.

The Open Source community and the commercial software vendors (specifically Accelrys, ACD/Labs, GGA Software Inc., OpenEye Software Inc. and Dotmatics Limited) are acknowledged for the valuable contributions their software has made to the development of our software. The active participation of many contributors, especially curators, data depositors and our users has greatly assisted us in expanding the quantity and quality of data. Their efforts are enabling improved science.

References

1. Williams, A. J. A perspective of publicly accessible/open-access chemistry databases. *Drug Discovery Today* **2008**, *13* (11−12), 495–501.
2. Williams, A. J. Public chemical compound databases. *Curr. Opin. Drug Discovery Dev.* **2008**, *11* (3), 393–404.
3. Williams, A. J.; Ekins, S. A quality alert and call for improved curation of public chemistry databases. *Drug Discovery Today* **2011**, *16*, 747–750.
4. Williams, A. J.; Ekins, S.; Tkachenko, V. Towards a gold standard: Regarding quality in public domain chemistry databases and approaches to improving the situation. *Drug Discovery Today* **2012**.
5. Pence, H.; Williams, A. J. ChemSpider: An online chemical information resource. *J. Chem. Educ.* **2010**, *87* (11), 1123–1124.
6. Williams, A. J., ChemSpider: Integrating Structure-Based Resources Distributed across the Internet. In *Enhancing Learning with Online Resources, Social Networking, and Digital Libraries*; Belford, R., Moore, J. W., Pence, H. E., Eds.; ACS Symposium Sereis 1060; American Chemical Society: Washington, DC, 2010; pp 23−29.
7. ChemSpider. http://www.chemspider.com (accessed June 2012).

8. Williams, A. J.; Tkachenko, V.; Batchelor, C.; Day, A.; Kidd, R., Utilizing Open Source Software To Facilitate Communication of Chemistry at the Royal Society of Chemistry. In *Free and Open Source Software in Applied Life Science and Industry*; Harland, L., Forster, M. J., Eds.; Biohealthcare Publishing Limited: Oxford, 2012.
9. Little, J. L.; Williams, A. J.; Pshenichnov, A.; Tkachenko, V. Identification of "known unknowns" utilizing accurate mass data and ChemSpider. *J. Am. Soc. Mass Spectrom.* **2012**, *23* (1), 179–185.
10. Hettne, K. M.; Williams, A. J.; van Mulligen, E. M.; Kleinjans, J.; Tkachenko, V.; Kors, J. A. Automatic vs. manual curation of a multi-source chemical dictionary: The impact on text mining. *J. Cheminf.* **2010**, *2* (1), 4.
11. Project Prospect Wins ALPSP Award. http://www.rsc.org/Publishing/Journals/News/ALPSP_2007_award.asp (accessed September 2011).
12. PubMed. http://www.ncbi.nlm.nih.gov/pubmed/ (accessed June 2012).
13. Royal Society of Chemistry Acquires ChemSpider. http://www.rsc.org/AboutUs/News/PressReleases/2009/ChemSpider.asp (accessed June 2012).
14. ChemSpider SyntheticPages. http://cssp.chemspider.com.
15. CAS SciFinder Database. http://www.cas.org/products/scifindr/index.html (accessed June 2012).
16. LearnChemistry. http://www.rsc.org/learnchemistry (accessed June 2012).
17. SpectraSchool. http://spectraschool.rsc.org/ (accessed May 2012).
18. Bradley, J. C.; Lancashire, R. J.; Lang, A. S.; Williams, A. J. The Spectral Game: Leveraging Open Data and crowdsourcing for education. *J. Cheminf.* **2009**, *1* (1), 9.

Chapter 3

Supplemental Journal Article Materials

David P. Martinsen*

American Chemical Society, 1155 16th St. NW, Washington, DC 20036
*E-mail: d_martinsen@acs.org

With the migration of journals from print to online environments over the last 15 years, many disciplines have seen an increase in content that is stored outside of the context of the traditional article. In the print world, the rationale for supplemental material was related to technical reasons (certain content could not be rendered in print) or economic reasons (the cost of processing or printing the content was prohibitive). In the digital world, it was assumed that the technical and economic reasons were, or would soon be, resolved. It was also assumed that there was value in including more data with an article in order to enable readers to fully understand the science. As journals saw increasing contributions of supplemental material, editors, reviewers, publishers, and readers experienced an impact from handling, reviewing, or reading the additional materials. NISO and NFAIS convened an exploratory group of publishers that led to the formation of a Working Group to develop best practice recommendations for publishing supplemental materials (NFAIS Supplemental Journal Article Materials Project. http://www.niso.org/workrooms/supplemental). While supplemental materials may contain primary research data, they also can contain much more. Accordingly, the NISO/NFAIS focus is broader than primary data.

Introduction

The current focus on primary research data is causing pressures on two extremes. On the one hand, some publishers are trying to rein in the quantity of data being submitted with research articles because of the increasing load on editors, peer reviewers, and readers. On the other hand, funding agencies are

© 2012 American Chemical Society

exploring ways to increase the preservation and accessibility of primary research data to increase the impact of the research they support. Several different types of organizations are trying to address the problem, including individual institutions, consortia, publishers, vendors, and government agencies.

The pressures on publishers to handle supplemental materials in the post-print era resulted in the formation of a Working Group under the auspices of the National Information Standards Organization (NISO) and the National Federation of Advanced Information Services (NFAIS) (*1*). A short survey sent to the CrossRef Technical Working Group and eXstyles listserves by Alexander ('Sasha') Schwarzman, then at the American Geophysical Union, revealed that there were very different practices across scientific, technical, and medical (STM) publishers in defining, processing, and disseminating supplemental materials (*2*). As a result, a meeting of interested parties was held at the American Psychological Association in January 2010, to see if there would be interest in establishing a Working Group to address the matter. Over 50 people representing 30 different organizations attended either in person or by phone (*3*). The group decided there was a need for establishing some best practices for supplemental materials, so that authors, readers, editors, publishers, and librarians might have some common expectations across journals and across publishers, or at least a way to determine the practices for a given journal. As a result, the NISO/NFAIS Working Group on Supplemental Journal Article Materials was formed. There were actually two Working Groups created, with the mission to generate best practice recommendations (as opposed to standards) for the publication of supplemental materials. The Business Working Group (BWG) was charged with making recommendations about policy and business questions. The Technical Working Group (TWG) objective was to make technical recommendations to enable the recommendations from the BWG to be implemented. As of this writing, the BWG Draft Recommendation has been through the public comment periods and comments have been incorporated into the draft. The TWG recommendation was released for public comment in July, 2012. Since the public comment period is still open, any comments on the NISO/NFAIS recommendations are subject to change. The NISO website should be referenced for the latest details.

Even before the Working Group was formed, Emilie Marcus, CEO and Editor of *Cell*, pointed out the growth in supplemental material in *Cell* and took steps to exert some control over what could be submitted as supplemental material (*4*). During the course of the discussions of the Working Group, some additional changes took place in the STM publishing world. First, the *Journal of Neuroscience* decided to stop accepting supplemental materials for publication with their journal articles (*5*). The reasons cited by the journal and approved by the Society of Neuroscience Council were the burden on reviewers and editors, and the perception that the journal was creating a collection of material that was only tangential to the articles. The *Journal* did not begin to publish supplemental material until 2003, but had seen a rapid increase in material since that time. As an alternative, scientists were encouraged to deposit their research data in suitable disciplinary repositories; the material would no longer be reviewed during the manuscript review. Under the new policy, reviewers and editors were not allowed to ask authors for supplemental material; one of the reasons authors were submitting more supplemental materials was as a pre-emptive measure, so that requests from editors or reviewers for additional materials would not slow down the review and publication process. In a similar vein, the *Journal of Experimental Medicine* made the decision to accept only "essential supporting information (6)."

On the other hand, in 2001, the International Union of Crystallography began to publish *Acta Crystallographica Section E: Structure Reports Online* (7). This is a journal that publishes articles consisting of crystallography data, an abstract, and perhaps a comment section. It is what today might be referred to as a data journal. Recent launches of *GigaScience* (8) and the *Geoscience Data Journal* (9) are also examples of data journals that are embracing data publication.

Another change that was announced during the Working Group discussions was that the *Biophysical Journal* began to ask that references cited from within the supplemental materials should also be included within the article (10). In the view of the *Biophysical Journal*, cited references in the supporting material were considered to be of equal importance to those in the main text, and should be included in impact factor and h-index calculations. Including the supporting references in the main text would ensure that they would be picked up by the citation services.

In the early days of migration to Web-based journals, there was a view that without the limitations of print, the article would migrate to something that was much more dynamic. This was certainly true in chemistry, where it was envisioned that the raw data could be included rather than bitmapped graphic images of structures and data. That has happened to some degree, but the current state of online journals has not reached the early vision, in part because of reader preference for the PDF, and in part because authors and publishers haven't solved the problem of how to capture, store, and deliver the content in a well-integrated fashion. Several projects have sought to envision the "article of the future", with *Cell* (11) and the Optical Society's Interactive Science Publishing (12) being recent examples. Earlier examples from the chemistry world include the *Internet Journal of Chemistry* (13) and the JUMBO exemplars (14) of the late 1990s.

With the fact that data, often marginally usable or useful data, could be uploaded and published relatively easily without additional curation on the part of the publisher, supplemental material did perhaps become a data dumping ground. Several publishers have documented a steep rise in supplemental material since 1996, the time when many journals began to publish on the Web, and began to receive manuscripts in digital formats instead of, or in addition to, paper.

A study by Smit and Gruttemeier discusses the increasing interest in the sharing of research data in an emerging data-intensive science (15). They look at the issues from the perspective of researchers themselves, who are more interested in accessing research data from other scientists than they are in sharing their own, but they also consider the perspectives of funding agencies, data-specific repositories, international organizations with a focus on data, and publishers. They provide some recommendations for publishers to consider, many of which are included in the NISO/NFAIS recommendations.

A Short History of the ACS Experience with Supplemental Materials

It is interesting to note though, that chemistry has had a somewhat different trajectory. It is possible that other fields have had a longer history with supplemental materials, but as far as the American Chemical Society (ACS) is concerned, inclusion of supplemental materials definitely pre-dates the Internet. A search of the ACS Publications website found a reference to supplemental materials as early as 1935, in the *Journal of Chemical Education (16)*, shown in Figure 1.

> **338**   JOURNAL OF CHEMICAL EDUCATION
>
> **COOPERATION WITH SCIENCE SERVICE**
>
> AUTHORS of manuscripts submitted to THIS JOURNAL who wish to publish related supplementary material or longer versions than those accepted by the editor may submit manuscripts embodying such material for approval. After this approval these manuscripts will be forwarded to *Science Service* for publication as *Science Service Documents* available in the form of microfilms or photoprints. Such manuscripts should be typewritten in an acceptable standard form (black fresh ribbon on 8½" X 11" bond paper, single spaced, preferably pica type) and should have any photographs or figures separately mounted on sheets of the same size.
>
> In these cases a footnote will be appended to the article as published stating that the more extended version or, as the case may be, the supplementary material is obtainable through *Science Service*. The cost, payable in advance by check or money order to *Science Service*, 2101 Constitution Ave., Washington, D. C., will be stated in the footnote. *Science Service Documents* in microfilm form (images 1 inch high on standard 35-mm. motion picture film) will cost approximately 1 cent a page; in photoprint form (6 X 8 inches in size, readable with the unaided eye) approximately 5 cents a page.
>
> *Science Service* also operates as a service to scientific research workers the *Bibliofilm Service* which copies to order literature in the library of the U. S. Department of Agriculture. Full information and order blanks for *Bibliofilm Service* and information about mechanisms for reading microfilms will be furnished free on request to *Science Service*. See also J. CHEM. EDUC., 12, 415-8 (1935).
>
>   OTTO REINMUTH, Editor

Figure 1. Announcement of a partnership with the Science Service for handling supplemental materials.

This advance in providing supplemental materials was made possible by a fascinating new technology called "filmstats". The *Journal of Chemical Education* had pronounced this technology a "new means for the advancement of science" (*17*). In a statement very similar to those of recent years, that article closes, "Many applications not apparent at present will certainly be found and it is reasonable to expect that film copying will gradually revolutionize the existing methods of distributing scientific and other recorded intelligence." In retrospect, filmstats were not quite as revolutionary as predicted, but they did provide a convenient method of distributing and perhaps more importantly, archiving, scientific intelligence. They also made possible the storage and distribution of supplemental materials in a more economic way than the conventional method of that time, which was paper. With respect to the ACS supplemental content, the Science Service, which had been housed at the National Academy of Sciences, eventually ended up at the Smithsonian Institution, where they still can be found today (*18*).

It is interesting to note that the Royal Society of Chemistry has an even earlier example of supplemental material. In a talk on the digitization of the RSC archive (*19*), Richard Kidd pointed to an article by Leeson in which a template for a physical device was included as supplemental material 1843 (*20*). Plates 1-6 of that article were not paginated; the intention was that the reader would cut out the template on each page, and connect them to create a working light polarization device. This supplemental material has been incorporated into the digital archive, currently as an Adobe Flash file (21).

Over the years, the ACS supplemental material collection moved around. After the arrangement with the Science Service, and between 1953 and the late 1960's, supplemental material was housed at the American Documentation Institute at the Library of Congress, as shown in Figure 2. This service was created to support supplemental materials from multiple publishers in order to best take advantage of the new technologies, microfilm and microfiche, which were emerging at that time. These materials are still available from the Technical Reports and Standards Division at the Library of Congress (22).

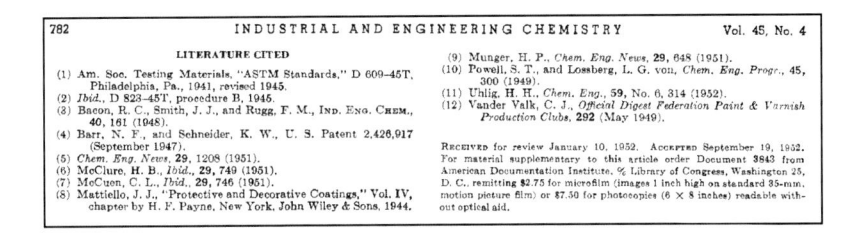

Figure 2. Supplemental Material paragraph referring to material stored at the Library of Congress (23).

In 1968, arrangements for ACS supplemental materials were moved to the American Society for Information Science, National Auxiliary Publications Service in New York, as shown in Figure 3. Through an arrangement with Microfiche Publications, ASIS/NAPS provided microfilm and microfiche services for publishers (24). Unfortunately, Microfiche Publications is no longer in existence, and ASIS, now ASIS&T, does not know what happened to those materials, pointing to the fact that attention to preservation predates the digital world.

Figure 3. Supplemental Material requests circa 1968-1971 (25).

By 1972, the ACS began to administer the supplemental materials from its Books and Journals offices in Washington, DC, as shown in Figure 4.

> (4) Listings of supplemental notes and material appearing in the microfilm edition of this volume of the journal are designated throughout the article with "s" notation, together with a brief comment on the nature of the supplemental material in brackets. Single copies may be obtained from the Business Operations Office, Books and Journals Division, American Chemical Society, 1155 Sixteenth St. N.W., Washington, D. C. 20036, by referring to the code number JPC-72-3603. Remit check or money order for $4.00 for photocopy or $2.00 for microfiche.
>
> (5) R. D. Evans, "The Atomic Nucleus," McGraw-Hill, New York, N. Y., 1962, Chapters 26–28. The text includes references to a number of the prior studies which "verified" the generality of the Poisson.
>
> *The Journal of Physical Chemistry, Vol. 76, No. 24, 1972*

Figure 4. A supplemental material reference referring the reader to the ACS Books and Journals offices in Washington, DC (26).

A 1964 note in the *Journal of Chemical Education* highlights another topic, which is behind much of the emerging concerns about preservation of primary research data (*27*), shown in Figure 5. A major concern at that time was fire; the article describes two incidents in which labs, libraries, research materials and data for Masters and Ph.D. theses, were destroyed. Fire is certainly still a cause for concern, but is probably outweighed by the loss of data from theft or loss of computers, or damage to computers or digital media. These latter problems are far more common than fire, but the loss of research data can be just as devastating.

Figure 6 shows a description from a 1941 article, and indicates that the reason for including supplemental material was to show the results of measurements, but which would have taken up too much space to include in the printed article.

Several years before launching its journals on the World Wide Web, the ACS began scanning supplemental materials for the *Journal of the American Chemical Society*, and posting those on a Gopher server (*29*). That procedure started in 1992, and expanded to include *Biochemistry* and *The Journal of Organic Chemistry* in 1994. The supplemental materials were also included in the CD-ROM versions of those journals. In mid-1995, with this migration of supplemental materials to a digital format, the supplemental materials were renamed to "Supporting Information". This was, at least in part, an indication that the scope of the content was expanding. In addition to content submitted on paper, digital files were also being accepted. This information included content having a greater degree of importance to the article.

These examples above show that while ACS has also seen an increase in articles containing supplemental materials, there has been a much longer history of supplemental materials than, for example, *The Journal of Neuroscience*. The move to supplemental materials was always seen as one way to keep page limits reasonable. By moving content from the main article in print to supplemental materials, where it is readily available online, the growth in the number of printed pages, and therefore the cost of editing, printing and shipping could be kept lower than would have been possible otherwise. The content was still considered significant to the article, but it was moved to the supplemental category to control costs.

> **PROTECTION OF RECORDS**
>
> "A one hundred thousand dollar piece of laboratory equipment would be easy to replace compared to the notes taken during certain experiments. Yet these notes and the valuable supplementary material, such as slides and samples, are often subjected to minimum fire protection; in fact, they are usually left on desk tops or in desk drawers which offer very little protection or none at all. Protection of records is an exacting study which is laid out in precise, workable form in NFPA Pamphlet No. 232. By following the suggestions in NFPA No. 232, appropriate economic protection for these records can be planned."
>
> *Jan. 23, 1958, Loretto, Pa., $280,000*
>
> "A fire thought to have been of incendiary origin, was discovered at 5:20 A.M. by an automatic fire alarm system in this 2-story and basement wooden college laboratory-library building that had no sprinkler system. The watchman had been through the building only 20 min before. The fire spread quickly over the wooden walls and up the open stairways. By the time the distant fire departments (6–18 miles away) arrived only the neighboring buildings could be protected. Destroyed along with a chemical lab, a biology lab, and a 15,000-volume library, were personal libraries, a master's thesis, 500 slides prepared for a doctor's thesis, two rare books printed before 1500. The school's previous 15,000-volume library was entirely destroyed ($250,000) in 1942 and in 1957 a $30,000 fire destroyed Giles Hall. None of these buildings had a sprinkler system and none had fire-resistive containers to protect records."
>
> *Feb. 5, 1956, Minneapolis, $500,000*
>
> "A fire of unknown origin was discovered by police in a patrol car as the flames became visible from the exterior of the laboratory building. The alarm was immediately radioed and two additional alarms were rung within the next 10 min. Fire had such a head start in this 20-year-old, 2-story wooden laboratory that only one wall and a portion of another were left standing at the end of the fire.
>
> A 20-year-old safe of 2-hr fire-resistive construction was severely damaged in its plunge to the cellar. The important records it contained were damaged by water as it sat submerged in the cellar hole until salvaged. Most of the records in this safe were, however, still legible.
>
> All other laboratory equipment, records, and engineering designs were destroyed by the fire. There were no sprinklers, watchman, or automatic detection within this combustible wooden building."
>
> *Vol. 41, No. 11, November 1964 / **A859***

Figure 5. Concerns about the protection of primary research data.

> given, because the other two gave much the same result.
>
> The two low molecular weight samples (for disks 677 and 678) were obtained by fractional precipitation of an acetone extract of L38. Four 100-g. batches were extracted with 2 liters of acetone
>
> probably reliable to better than 0.5%.
>
> In order to determine whether the thermal
>
> (12) Measurements were made at other temperatures, but in order to save space are omitted here. For a copy of these data, order Document 1561 from the American Documentation Institute, Offices of Science Service, 2101 Constitution Ave., Washington, D. C., remitting 23c for microfilm or 50c for photocopies readable without optical aid.

Figure 6. From a 1941 article in J. Am. Chem. Soc (28).

In addition to costs, though, there was also a focus on making the digital supporting information more useful. A survey of the author instructions for ACS journals reveals a number of recommendations for supporting information. For certain journals, specific data files are required to validate the identity and purity of new molecules or synthetic procedures. For example, *The Journal of Organic Chemistry* requires a compound characterization checklist in certain instances, and states the "Data needed to document structure assignments, purity assessments, and other conclusions should be included in the manuscript or in the supporting information (*30*)." In other journals, and for other types of experiments, authors are asked to submit data files to repositories where the data will be validated and made available to the community. In *Biochemistry*, for example, authors are asked to deposit data in the Protein Identification Resource, GenBank, the Protein Data Bank, the BioMagResBank, the Nucleic Acid Data Bank, and the BindingDB (*31*). These are just examples of the variety of considerations within different disciplines within chemistry.

The NISO/NFAIS Supplemental Journal Article Materials Working Group

One of the earliest challenges for the NISO/NFAIS Working Group came in the form of definitions. Before best practices recommendations could be made for supplemental materials, the group felt that there had to be a good definition of supplemental materials, and for that, there needed to be a definition of the article; that is, the entity that the supplemental materials supplemented. This proved to be a somewhat elusive concept. Part of the reason is that scholarly publishing is still in a transition period between the print and electronic worlds. A printed volume of bound pages, paginated sequentially, is still viewed by many as the collection of articles. PDF files, though now delivered digitally to the reader, are often printed locally and are equivalent to the unbound printed volume. Anything that could not be represented in print was considered supplemental. This is a fairly easy definition; the article is distinguished from the supplemental material by the medium or format in which it can be disseminated. However, a number of publishers now see the online version as the canonical version of an article; there may be interactive components or objects inserted into the HTML version. It is now possible to embed interactive objects into the PDF as well (*32*). These cannot be represented in print except by a static image.

Because the transition is not complete, there is still a general practice to designate material as supplemental based on format. However, this is a somewhat artificial distinction. Even in today's world, there are figures within the running text of an article and figures in supplemental material that are the same type of content and in the same format. The reason that some of these are considered part of the article, while some are considered supplemental, has more to do with the relevance of the content to the understanding of the article than the format. In the end, the definition of the article stressed that it was an original publication in a scholarly journal, and consisted of all of the materials necessary for a reader to comprehend the work.

Types of Supplemental Journal Article Materials

After much discussion, the BWG felt it would be useful to divide the supplemental materials into three categories. The first category is that which is fundamental to understanding the article, which was designated as integral.

At first glance, this may seem like a contradiction. How could something be supplemental and integral at the same time? If something is integral to the article, why is it not included within the article itself? For technology or economic reasons, it may not be feasible to include the integral supplemental material within the article. Note that the technological reasons could be a limiting factor on both the production side and the delivery side. For example, it might be possible to include the material within the article, but in doing so, it might make it difficult for a significant percentage of readers to read such an article. Also note that while this integral material is necessary for a full understanding of the science being reported, it might not be at the same level of detail required for full replication of the experiments.

When supplemental material is categorized as integral, the author, editor, reviewer, publisher, and reader should expect that material to have been peer reviewed at the same level as the article itself. Since the material is integral to the article, the best practice recommendation currently states that the presence of the supplemental material and the link to that material be included within the context of the article where it is being described or referenced. At the current time, integral supplemental material will usually be hosted at the publisher site. However, as noted above, publishers are establishing relationships with discipline-specific repositories for hosting certain types of data. In the absence of a formal relationship, repositories that have become standard in certain areas, such as the Protein Data Bank, are recommended as sites for depositing data either before, or concurrent with, the submission of a manuscript. In an online environment, links to data at these repositories can be included within the text of the article or in a sidebar for easy access by the reader. When supplemental material is hosted in a repository other than at the publisher site, it is important that the publisher consider the long term viability of the links, as well as of the repository, and develop a plan to ensure integral content is always available to future readers of the articles.

The second category of supplemental material was designated as additional. Additional supplemental material is not required to understand the science, but is useful to help the expert reader to understand the research at a deeper level, might provide more information for an expert to replicate the experiments, might give details on similar experiments to those reported in the article, and so on. It would be desirable if this material was also reviewed at the same level as the article, but since it is not critical material, such review is considered optional. The additional material may or may not be hosted on the publisher platform, or the same platform as the article. The link to the additional material should be included as a cited reference at the end of the article.

The third type of supplemental material was simply called related material. The nature of this material is somewhat open ended; most likely it will not be hosted by the publisher. The Working Group considered this material beyond the scope of the recommendations, since the publisher has little or no control over the content, and while persistent links to the related material are desirable, such links may or may not exist. The only recommendation here is that the material be hosted in a repository which has some commitment to preservation, more so than would be the case for the average author's personal website.

Discoverability of Supplemental Materials: Linking

One of the problems that the Working Group sought to address was related to discoverability of supplemental materials. While metadata practices and persistent identifiers associated with a journal article are well established, for example the National Library of Medicine (NLM) Journal Article Tagset (33), such is not the case with supplemental materials. It is often difficult for a reader to know whether or not supplemental materials exist for a given article. When a Web-based search leads directly to the supplemental material for an article, it may be difficult to locate the associated article. In order to address this problem, the Working Group recommended that the existence of supplemental material be highlighted in several places. These include on the Table of Contents listing for an article as well as in a prominent place on abstract, html, and PDF views of article text. Machine-readable elements indicating the existence of supplemental material would also be helpful for identifying which articles contain supplemental material. The NLM Journal Article DTD (document type definition) already has elements for declaring that supplemental material exists. When using the NLM DTD, these elements should be used.

Supplemental materials are often posted with little technical editing on the part of the publisher. However, in order to enable the relationship between the supplemental materials and the associated article, a set of metadata has been proposed for the supplemental material itself. One purpose is to provide the identity and the location of the article. This would allow the reader stumbling across the supplemental material to readily link back to the article, something that is not always possible today.

The Working Group recommends assignment of persistent identifiers to the supplemental material. For publishers of journal articles, this would most likely be a CrossRef-based DOI (34). However, with an emerging focus on storing and disseminating the primary research data supported by funding agencies and collected in academic laboratories, DataCite was formed to develop and encourage best practices for that domain (35). In some cases it might be appropriate for the supplemental material, particularly supplemental datasets, to be assigned a DataCite DOI that would be used to link from the journal article to the supplemental material. The Working Group also recognized that there are some well-established identifiers in certain disciplines, and these could also be used to link from a journal article to its supplemental material.

Peer Review of Supplemental Materials

One of the areas of uncertainty with supplemental material is whether or not the material has been subject to peer review. There is a wide variability in practice across publishers, from those who do not expect supplemental materials to be reviewed at all, to those who expect reviewers to consider the material in the same way they do the article. The recommendation of the Working Group is that publishers and editors consider that integral supplemental material be treated in the same way as the article itself. There is no recommendation regarding tagging the supplemental material as reviewed. However, the recent CrossMark (36) specification includes an element for a publisher to indicate that an article has been subjected to peer review. If CrossMark is widely adopted, the designation that materials have been reviewed may become more common in the future.

Archiving and Preservation of Supplemental Materials

Archiving and preservation are perhaps the most complex part of the recommendation to put into practice. Early in the transition from print to digital, many were concerned about the long term preservation of scholarly materials. In the print world, multiple copies of most journals could be found scattered institutions across the globe. While paper is subject to deterioration and is also somewhat fragile, it is not nearly as fragile as digital media. Even partially deteriorated paper might still be capable of being read by the human reader. Digital media might be considered as more ephemeral than paper because first of all the physical medium can deteriorate even when stored in climate controlled conditions. A partially damaged storage device might not be readable at all by a computer. Even a storage device that is not damaged at all may not be readable on future hardware.

In addition to preservation of the stored bits, there is another aspect concerning the ability to ingest the bits and render them in the same way as the author intended and that the original readers saw them. In the traditional world, where an article was made up of text and some static images, the requirement for rendering the bits was not complex. While PDF files, tiff images, mp3, mp4, and a host of other formats work fine most of the time, the future viability of these formats is not known. Given the large number of files in common formats, it is reasonable to expect that these formats will continue to be supported, or that conversion software will be written to convert them into newer formats. For less common formats, though, that expectation may not be justified. For this reason, the Working Group recommends that some preservation considerations be taken into account when publishers and editors determine what formats to accept as supplemental materials. For integral material, a strict adherence to the selected formats should be expected. For additional and other materials, the format restrictions can be lessened, for example, to allow the use of emerging technologies. Publishers are encouraged to make available a list of acceptable formats as well as applications that can be used to render those formats. Reference is made to two format repositories, currently under development, which store information related to file format (*37*). However, while these universal registries are being developed, most publishers will likely maintain a list of formats and applications on their own sites so users can easily find the necessary information to use the supplemental material.

It is worth noting that many different groups are investigating the question of preservation of scholarly research data, including funding agencies, universities, institutional libraries, and government agencies. The focus of the Working Group, and of this chapter, is on the publisher perspective on materials associated with journal articles. Some of those materials may be datasets, but datasets are not the primary focus. Readers are referred to other chapters in this volume that deal more broadly with storage and preservation of research outputs by others in the research community.

Dissemination of Supplemental Materials: Packaging

In order to better enable the association of a journal article with all of its relevant objects, the TWG considered how an article and its supplemental materials could best be bundled and transferred as a complete unit. As mentioned above, there is a fuzzy line between the components of the article and the components of the supplemental materials. They often include objects of the same nature and the same format, with only a labeling difference between them, e.g., "Figure 1" vs. "Figure S1". A recommendation on how to include all of the components of an article was beyond the scope of the Working Group; nonetheless, the Group felt that it was important to consider this. The recommendation in this area focuses on a manifest listing all of the files associated with an article, with all files grouped within a container, for example a .zip file. That recommendation is still under discussion at the time of this writing.

Acknowledgments

The author wishes to thank all of the members of the NISO/NFAIS Business and Technical Working Groups for the interesting and productive discussions during the development of the recommendations for supplemental journal article materials, especially the co-chairs of the BWG (Linda Beebe of the American Psychological Association and Marie McVeigh of Thomson Reuters) and my co-chair on the TWG (Alexander ('Sasha') Schwarzman of the American Geophysical Union, and now at the Optical Society). The full list of Working Group members can be found at:

http://www.niso.org/workrooms/supplemental/technical/
http://www.niso.org/workrooms/supplemental/business/

References

1. NFAIS Supplemental Journal Article Materials Project. http://www.niso.org/workrooms/supplemental.
2. Schwarzman, A. Supporting Material, November 2, 2009. http://www.agu.org/dtd/Presentations/sup-mat/sup-mat.pdf (accessed May 25, 2012).
3. A report on the meeting can be found on the NISO Web site at http://www.niso.org/apps/group_public/download.php/3708/NFAIS_NISO_Supp_Materials_Meeting_Summary_Report_rev.pdf.
4. Marcus, E. *Cell*, **2009**, *139* (1), 11. http://10.1016/j.cell.2009.09.021.
5. Maunsell, J. Announcement regarding supplemental material. *J. Neurosci.* **2010**, *30* (32), 10599–1060011 August 2010.
6. Borowski, C. Enough is enough, *J. Exp. Med.* **2011**, *208* (7), 1337. http://dx.doi.org/10.1084/jem.20111061.
7. Clegg, W.; Watson, D. G. Structure reports online: The birth of a new journal. *Acta Cryst. E* **2001**, *57* (1), e1-e2. http://dx.doi.org/10.1107/S1600536800020432.

8. Goodman, L.; Edmunds, S. C.; Basford, A. T. Large and linked in scientific publishing. *GigaScience* **2012**, *1* (1), 1−2. http://dx.doi.org/ 10.1186/2047-217X-1-1.
9. Geoscience Data Journal. http://onlinelibrary.wiley.com/journal/10.1002/(ISSN)2049-6060 (accessed July 31, 2012).
10. Biophysical Journal. Author Guidelines: Supporting Material, June 21, 2011. http://download.cell.com/images/edimages/Biophys/Supporting_Material.pdf (downloaded May 8, 2012).
11. Marcus, E. 2010: A publishing odyssey. *Cell* **2010**, *140* (1), 9. http://dx.doi.org/10.1016/j.cell.2009.12.048.
12. See for example Hong, J.; Kim, Y.; Choi, H.; Hahn, J.; Park, J.; Kim, H.; Min, S.; Chen, N.; Lee, B. Three-dimensional display technologies of recent interest: Principles, status, and issues. *Appl. Opt.* **2011**, *50*, H87−H115. http://dx.doi.org/10.1364/AO.50.000H87.
13. Bachrach, S. M.; Heller, S. R. The internet journal of chemistry: A case study of an electronic chemistry journal. *Serials Review* **2000**, *26* (2), 3−14. http://dx.doi.org/10.1016/S0098-7913(00)00054-X.
14. Rzepa, H. S.; Murray-Rust, P.; Whitaker, B. J. The application of chemical multipurpose internet mail extensions (Chemical MIME) Internet standards to electronic mail and World Wide Web information exchange, *J. Chem. Inf. Model.* **1998**, *38* (6), 976−982. http://dx.doi.org/10.1021/ci9803233.
15. Smit, E.; Gruttemeier, H Are scholarly publications ready for the data era? Suggestions for best practice guidelines and common standards for the integration of data and publications. *New Review Information Networking*, **2011**, *16* (1), 54−70. http://dx.doi.org/10.1080/13614576.2011.574488.
16. Reinmuth, O. Cooperation with science service. *J. Chem. Educ.* **1936**, *13* (7), 338. http://dx.doi.org/10.1021/ed013p338.1.
17. Seidell, A. Filmstats, a new means for the advancement of science. *J. Chem. Educ.* **1935**, *12* (9), 415–41810.1021/ed012p415.
18. SIA RU007091, Science Service, Records, circa 1910−1963. http://siarchives.si.edu/collections/siris_arc_217249 (accessed May 5, 2012).
19. Kidd, R. Publishing Innovation at the Royal Society of Chemistry, 230th ACS National Meeting, Washington, DC, August 28−September 1, 2005, CINF-56.
20. Leesen, H. B. XCI. Observations on the circular polarization of light by transmission through fluids. *Mem. Proc. Chem. Soc.* **1843**, 2, 26−45. http://dx.doi.org/10.1039/MP8430200026.
21. Memoirs and Proceedings of the Chemical Society, Interactive Plates from Volumes II and III. http://www.rsc.org/Publishing/Journals/DigitalArchive/InteractivePlates.asp (accessed May 5, 2012).
22. Information about the Library of Congress Technical Reports and Standards Division can be found at http://www.loc.gov/rr/scitech/trs/trsover.html (accessed May 24, 2012). Details about the collection can be found at http://www.loc.gov/rr/scitech/trs/trsadi.html (accessed May 24, 2012).
23. Massa, A. P.; Colon, H.; Schurig, W. F. *Ind. Eng. Chem.* **1953**, *45* (4), 775–782. http://dx.doi.org/10.1021/ie50520a034.

24. Notices *Am. Miner.* **1973**, *58*, 570−572. http://www.minsocam.org/ammin/AM58/AM58_570.pdf (accessed May 24, 2012).
25. Ritchie, A. W.; Nixon, A. C. Dehydrogenation of monocyclic naphthenes over a platinum on alumina catalyst without added hydrogen. *Ind. Eng. Chem. Prod. Res. Dev.* **1968**, *7* (3), 209–215. http://dx.doi.org/10.1021/i360027a011.
26. Anderson, J. L. Non-Poisson distributions observed during counting of certain carbon-14-labeled organic (sub)monolayers. *J. Phys. Chem.* **1972**, *76* (24), 3603–3612. http://dx.doi.org/10.1021/j100668a018.
27. Steere, N. V. Fire-protected storage for records and chemicals. *J. Chem. Educ.* **1964**, *41* (11), p A859. http://dx.doi.org/10.1021/ed041pA859.
28. Fuoss, R. M. Electrical properties of solids. IX.1 Dependence of dispersion on molecular weight in the system polyvinyl chloride-diphenyl. *J. Am. Chem. Soc.* **1941**, *63* (9), 2401–2409. http://dx.doi.org/10.1021/ja01854a026.
29. Gopher was a protocol developed at the University of Minnesota about 1991 to transfer files over the Internet. It was directory, file, and text based, and was easily implemented. It was popular for several years before being eclipsed by the http protocol (which had been introduced about the same time). See the Wikipedia entry for more details: http://en.wikipedia.org/wiki/Gopher_(protocol) (accessed August 12012).
30. The Journal of Organic Chemistry, Guidelines for Authors, Updated January 2012. http://pubs.acs.org/paragonplus/submission/joceah/joceah_authguide.pdf (accessed May 25, 2012).
31. Biochemistry, Author Guidelines (Revised January 2012). http://pubs.acs.org/paragonplus/submission/bichaw/bichaw_authguide.pdf (accessed May 25, 2012).
32. Selvam, L.; Vasilyev, V.; Wang, F. Methylation of zebularine: A quantum mechanical study incorporating interactive 3D PDF graphs. *J. Phys. Chem. B* **2009**, *113* (33), 11496−11504. http://dx.doi.org/10.1021/jp901678g.
33. See for example the NLM Journal Article Tag Set http://jats.nlm.nih.gov/index.html and the NISO Z39-96 Draft Standard http://www.niso.org/standards/z39-96/ (accessed May 25, 2012).
34. Note that CrossRef is simply one of many DOI Registration Agencies and assigns DOIs to the scholarly literature. Other Registration Agencies focus on other types of materials or with non-English language materials. A list of DOI Registration Agencies can be found at http://www.doi.org/registration_agencies.html.
35. DataCite is a collaboration between academic institutions and national libraries to develop an infrastructure to store metadata and persistent identifiers to datasets. At least one of the goals is to encourage citation of datasets as a means to give credit to a research output which is not recognized through authorship on the published article, but which has some merit in itself. See http://www.datacite.org for additional information.
36. CrossMark is a protocol to enable publishers to designate a version-of-record for an article, and also to allow readers to determine whether or not they

have the most up-to-date version of an article. Details can be found at http://www.crossref.org/crossmark/index.html (accessed May 5, 2012).

37. The Unified Digital Formats Registry (UDFR) was developed by the University of California Curation Center and the California Digital Library, with funding from the Library of Congress (http://www.udfr.org/, accessed May 5, 2012). This registry grew out of the earlier Global Digital Format Registry, developed by Harvard University, OCLC, and the National Archives and Records Administration with Mellon Foundation funding (http://gdfr.info/, accessed May 5, 2012), and the UK National Archives PRONOM program (http://www.nationalarchives.gov.uk/PRONOM/Default.aspx, accessed May 5, 2012). While the GDFR program has is no longer active, the PRONOM effort is ongoing.

Chapter 4

National Data Management Initiatives and the U.S. Exemplar: DataONE

Suzie Allard*

School of Information Sciences, University of Tennessee,
453 Communication Building, Knoxville, Tennessee 37996-0341
*E-mail: sallard@utk.edu

This chapter discusses data management initiatives that are emerging around the world. It briefly reviews national level initiatives that are growing in Australia, Germany, The Netherlands, the United Kingdom and the United States since these initiatives are establishing a foundation for other countries. The chapter then introduces the United States' National Science Foundation's DataNet initiative and looks at an exemplar project – DataONE. DataONE is an NSF-funded project that focuses on creating sustainable cyberinfrastructure in the domain of environmental science.

Data are the lifeblood of research. Trusted data, data which are verifiable and persistent, are essential to producing meaningful results that expand our knowledge and can be used for encouraging innovation and making informed decisions for society. Data that are sound, accessible, and persistent rely on good data management and curation practices to occur at all points in the data lifecycle. The need for data management has been identified as an important issue across many disciplines ranging from the humanities, to social sciences, to science. Researchers, information professionals, and agencies in countries across the globe recognize the need for data management. Many nations are introducing initiatives to encourage the development and adoption of effective data management strategies.

Science is at the nexus of a paradigmatic shift brought about by a rapidly expanding volume of data that can be collected automatically, increasing computational power for modeling and simulation, and improving technology

© 2012 American Chemical Society

for collaboration (*1*). This move to data-intensive scientific discovery has been called the fourth paradigm (*2*). The first scientific paradigm began more than a thousand years ago with the empirical description of natural phenomena. Science moved into the second paradigm only a few hundred years ago, with the advent of using models and generalizations. Computational capabilities that enabled simulations are the hallmark of the third paradigm. The fourth paradigm unifies theory, experiment, and simulation, as data are explored in new ways, changing how science is conducted (*3*) and also how the scholarly record is engaged by scientists and publishers (*4*).

An added incentive for developing initiatives is the recognition of the value of "Big Data." Big Data is the term used to describe data sets that are large, diverse, and complex which may be generated by instruments, sensors, internet transactions or other digital sources (*5*). Big Data require special tools to collect, store, analyze, visualize and share the data (*5*). Big Data analyses have led to new discoveries in fields ranging from environmental science to economics. In March 2012 in the United States, the Obama administration announced the "Big Data Research and Development Initiative" that dedicated $200 million to "improve the tools and techniques needed to access, organize, and glean discoveries from huge volumes of data" (*6*). The initiative will support the development of core technologies to enable big data collection, storage, preservation, analysis, and sharing, in order to accelerate discovery in science and engineering, to transform teaching and learning, and to expand the work force to address Big Data issues (*6*).

This chapter briefly introduces a selection of national data management initiatives from around the world listed in alphabetical order. These initiatives were chosen for this chapter because they were among the first to address the need for data management at a national level. It should be noted that there are other countries also working on their national information infrastructure (e.g. South Korea). However, the countries chosen for this chapter were selected based because the work they have accomplished in the early years of data management programs have made them prominent players in the world-wide data management community. After setting the global context of data management, the chapter's focus narrows to review the history of the United States' National Science Foundation's DataNet initiative. This program funds the exemplar cited in this chapter, DataONE. DataONE is focused on creating sustainable cyberinfrastructure in the domain of environmental science. DataONE's public release was in July 2012 and can be accessed at www.dataone.org.

Australia

Australia is actively embracing the need for data management, and aggressively crafting a strategy to implement an initiative to protect the digital data created by the country's researchers. The Australian National Data Service (ANDS) is a vibrant program that expresses its vision as "More Australian researchers reusing research data more often" (*7*). ANDS, which was established in 2008, is working to create structured data collections that are findable, well-managed, and reusable. ANDS is funded as part of the National

Collaborative Research Infrastructure Strategy (8) . This program notes that eResearch infrastructure (9) is an important component for Australian researchers. ANDS activities also receive some funding from the Education Investment Fund to benefit the establishment of the Australian Research Data Commons created by the Super Science Initiative that awarded $1.1 billion (Australian) to enhance the research infrastructure for science (10).

ANDS' strategy for improving data management and curation includes: (1) developing partnerships and collaborations with researchers and data-producing agencies; (2) providing national services; (3) providing guidance and materials about data management including reuse; (4) developing communities of practice; and (5) building the Australian Research Data Commons (11). ANDS funds projects related to data description and metadata, research data and public sector data management, appropriate tools, and community creation (11).

Improving how research data is managed in Australia is a priority for ANDS. ANDS has created a suite of resources for institutions, including information about creating a data management framework, defining research data, managing metadata, building research data policy, licensing data, and implementing the full data curation continuum. Another resource is a toolkit to use for determining the characteristics of existing data collections.

ANDS is reaching out to both institutions and individual researchers while promoting research to explore data management related challenges. Compared to the programs in other nations, it is still quite young, however it is making strides within the Australian research community.

Germany

Germany recognized the importance of protecting the scientific process nearly three decades ago, and, the focus on digital data and persistent access began about a decade ago. The German Research Foundation led the activities for a cohesive data management plan, including making a substantial investment in funding research related to data management.

The Deutsche Forschungsgemeinschaft (DFG) is the German Research Foundation that was founded in 1951 in the Federal Republic of Germany, and later extended to serve unified Germany in 1990 (12). The DFG focuses on funding science research, including improving the research infrastructure. This focus on the infrastructure includes the goal that research data should be freely and easily accessible and "should be professionally curated on a long-term basis" (13). The DFG participates in the Alliance of German Science Organizations on many issues, including data management. Participants include the Max-Planck Society, Helmholtz Association, Leibniz Association, and Frauenhoffer Association.

DFG began engaging data management issues in 1998 with the program, "Proposals for Safeguarding Good Scientific Practice." In 2003 the Max Planck Society hosted a meeting for scholars from many nations to explore the issues surrounding open access. The resulting Berlin Declaration on "Open Access to Knowledge in the Sciences and Humanities" proclaimed that disseminating knowledge was incomplete until the information is made easily available to society (*14*). The Berlin Declaration asked signatories to encourage researchers to embrace the open access paradigm while still maintaining good scientific practice for publication (*14*) and open access contributions specifically included "original scientific research results, raw data and metadata, sources materials, digital representations of pictorial and graphical materials and scholarly multimedia material" (*14*). In 2007, the DFG began sponsoring workshops to shape the national initiative, "Digital Information" (*13*). In 2010 the "Future of Information Infrastructure (KII)" began working on licensing, research data and virtual research environments, with a particular focus on data re-use (Fournier, 2011).

The DFG has been instrumental in leading many data management-related initiatives, including 2009's "Recommendations for Secure Storage and Availability of Digital Primary Research Data," which defined research data and metadata standards, established the rights of scientists and their access to data, established availability guidelines, and outlined quality control (*15*).

DFG's strategy in promoting good data management practices focuses on six key areas: (1) building awareness of how to use resources efficiently to make data available; (2) working closely with scientists who are regarded as both data producers and users; (3) determining unique discipline-specific needs; (4) developing a process for research data publication; (5) identifying best practices; and (6) supporting pilot projects. The "Information Infrastructures for Research Data" call in 2010 elicited 90 proposals, and funding awards to support this strategy were distributed to 27 projects representing research in the domains of life sciences, humanities, social sciences, natural sciences, and engineering. These projects have a range of goals, including creating disciplinary data centers (including some with Big Data), creating systems for linking data with publications, developing systems for persistent identifiers, developing workflow tools, and creating systems for long-term storage (*13*).

Germany's national level efforts are aimed at both the individual scientists and the scholarly process that must be implemented to support a successful data management initiative. The significant monetary support for projects that will help establish a national data infrastructure demonstrates Germany's commitment to this initiative.

The Netherlands

The Netherlands is working diligently on a national level initiative for data management. Their strategy includes directly addressing a range of stakeholders that includes researchers, data librarians and policy makers. The well-coordinated efforts are also reaching out to surrounding countries to help assure a multinational level of success.

The SURF Foundation and the Data Archiving and Networked Services (DANS) lead data management initiatives in The Netherlands. Established in 1987, SURF (Samenwerkende Universitaire Reken Faciliteiten), brings together Dutch higher education and research institutions in a partnership to advance the use of information and communication technology (*16*). One of the SURF focus areas is scholarly communication, including access to research data (*17*).

SURF identifies three primary stakeholder groups with whom they are working – researchers, data librarians, and policymakers. SURF also identified questions of interest to each of these groups, and the projects that are conducting research that addresses those questions. For example, SURF notes that the CARDS project addresses the question of how to support researchers (17). SURF includes an action plan to create a collaborative data infrastructure between four countries – Denmark, Germany, the Netherlands, and the United Kingdom (17).

DANS (18) was established in 2005 as an institute of the Royal Netherlands Academy of Arts and Sciences (KNAW) and is also supported by the Netherlands Academy of Arts and Sciences (NOW). DANS is focused on storing data from the arts, humanities, and social sciences, and providing persistent access to these data sets (19). DANS offers a range of services to support access to digital data. EASY is an online archiving system that supports persistent access to data and encourages researchers to archive and reuse data. NARCIS.nl is a website that provides access to data in a wide range of areas, including extensive data sets in the social sciences, history, and archeology (20). DANS also provides training and guidance for data producers and data users, as well as conducting research to explore solutions for sustained access to digital data (21). DANS has also created a Data Seal of Approval that is granted to data repositories that meet specific criteria that assure the quality, the preservation, and the long-term accessibility of the data.

The Netherlands' vision of making a multinational collaborative data infrastructure a key part of the initial design is unique. This illustrates the need for strategic thinking that answers a nation's specific needs when designing and implementing a national data initiative. This may need to be a strategic consideration for more national initiatives since many research question are global by nature and a data infrastructure that is collaborative may facilitate interoperability which could increase successful data sharing.

United Kingdom

The United Kingdom has been formally addressing issues of digital curation for nearly two decades by establishing the Digital Curation Centre. The UK program has been actively collaborating with other countries to share expertise and to align efforts as efficiently as possible.

The U.K.'s Digital Curation Centre's (DCC) motto, "because good research needs good data," eloquently summarizes the reason for supporting strong data curation and data management practice (22). The DCC, which serves the higher education research community, promotes "digital information curation" by focusing on "building capacity, capability, and skills for research data management" (22). The DCC's activities include provision of how-to guides and training programs to facilitate good data management practices among researchers and information professionals.

The DCC was established in March 2004 by a consortium of universities and other agencies. This consortium's founding members were the Universities of Edinburgh and Glasgow which together host the National eScience Centre, UKOLN at the University of Bath, and the Science and Technology Facilities

Council. The DCC's creation was based on the recommendation from a report, *Continuing Access and Digital Preservation Strategy,* produced by the Joint Information Systems Committee (JISC) (*23*). The JISC was created in 1993 to provide universities with the vision and guidance to adopt and use new technologies (*24*). JISC's role has changed as technologies have emerged and evolved. In late 2010, an independent review of JISC led to the restructuring of the organization so it could move more dynamically into the future. The restructuring to move towards a new, separate legal entity included identifying the best governance and business models, reviewing all JISC funded services, reviewing the operations at the Joint Academic Network (JANET) and gaining an understanding of the market that JISC will serve in the future. This transition is projected to complete in August 2012 (*25*).

The DCC has evolved in three phases. The first two phases (March 2004-Feb 2007, and March 2007-February 2010) focused on stakeholders who were involved in digital preservation and curation, such as data specialists, librarians, and archivists. While researchers and policy-makers were part of the initial outreach, activities involving direct interaction with the research community increased at the start of Phase 2. This included creating an e-Science Liaison (*26*) and conducting the Disciplinary Approaches to Sharing, Curation, Re-use, and Preservation (SCARP) case studies that illuminated the similarities and differences in data practices across different disciplines (*27*).

Phase 3 began in March of 2010 and the focus shifted to outreach to the 100,000 researchers in the U.K. by providing support to data custodians through DCC training programs. The idea is that these data custodians will then reach out to researchers who are both data producers and data users. The DCC believes that many researchers recognize that data management is important, but they do not yet have the tools and resources to implement good data management practices efficiently and effectively in their own work (*26*). The DCC was also reorganized into a consortium with the principal partners of the University of Edinburgh, the Humanities Advanced Technology and Information Institute (HATII) at the University of Glasgow and UKOLN (formerly known as The United Kingdom Office for Library and Information Networking) based at the University of Bath.

The DCC maintains relationships with organizations throughout the U.K. and around the world that are engaged in data curation and data management initiatives. These include organizations in Europe, Australia, and the United States – some of that are introduced below.

The United Kingdom's DCC has been at the forefront of digital curation and data management activities. The recent strategic shift to more directly reach out to UK researchers demonstrates how an initiative may mature over an extended period.

United States

Many federal agencies and other organizations in the United States are aggressively working to assure that research data will be preserved and persistently accessible. Many of these organizations have a science focus include, alphabetically, the Department of Defense, the National Aeronautics and Space Agency (NASA), the National Institutes of Health (NIH), National Oceanic and Atmospheric Administration (NOAA), the National Science Board, (NSB), and the United States Geological Survey (USGS). Interest in research data issues also extends to organizations that address domains other than science including, in alphabetical order, the Institute for Museum and Library Services (IMLS), the Library of Congress, the National Archives and Records Administration (NARA), and the National Endowment for the Humanities (NEH).

The organizations have different approaches to the mission, but their activities dovetail well to create an environment that fosters the emergence of other bodies focused on data management. This chapter will focus on two organizations that provide two models for working towards accessible and well-preserved data. The Coalition for Networked Information (CNI) is a membership-funded organization that provides vision and leadership across all domains and to a range of stakeholders but has limited funding so it cannot provide monetary support to specific projects. The National Science Foundation, a U.S. government funded agency, also provides vision and leadership to the scientific community and has the ability to support projects that can further this vision and make it reality.

The Coalition for Networked Information (CNI), established in 1990, is an organization born out of a joint initiative of the Association of Research Libraries and EDUCAUSE. It is funded by the membership dues of the more than 200 members including libraries and library organizations, institutions of higher education, and members in the fields of publishing and information technology as well as scholarly and professional societies. CNI focuses on advancing education and scholarship by promoting the adoption and use of enabling digital information technology (28).

One focus for CNI has been e-Research, including the issues surrounding data sets and data management in all parts of the data lifecycle. CNI helps facilitate better use of information technology to enhance collaborations between libraries, publishers, researchers, and others (29). CNI builds awareness of emerging trends and technologies at its semi-annual membership meetings that feature introductions to cutting edge projects that are changing the way scholarship and education is being conducted. For example, at the Spring 2012 Membership Meeting, there were talks and a focus group discussing data management in policy and education (30). While CNI is not an agency that provides funding to projects, its role as a thought-leader that serves as a catalyst for the broader community has been important.

The National Science Foundation is an agency created by the U.S. Congress in 1950, "to promote the progress of science…" (31). The NSF mandate is very broad; however, it has long had a focus on technologies that enhance science, starting in the 1960s with campus-based computing facilities, and including their support of super computing in the 1990s. NSF established the Office of Cyberinfrastructure (OCI) in the early 2000s to coordinate how advanced computational facilities can be used to solve complex problems facing the science and engineering disciplines. Cyberinfrastructure provides the technology and tools to support scientific inquiry (32) and includes both technological and sociological perspectives (National Science Foundation Blue-Ribbon

Panel on Cyberinfrastructure 2003). NSF presents the following definition of cyberinfrastructure: "the broad collection of computing systems, software, data acquisition and storage systems, and visualization environments, all generally linked by high-speed networks and often supported by expert professionals" (*33*).

In 2007, the OCI focused on developing cyberinfrastructure for digital research data by creating a new program, the Sustainable Digital Data Preservation and Access Network Partners, or DataNet (*34*). The DataNet program promoted multidisciplinary approaches to tackle data issues in order to:

"(1) provide reliable digital preservation, access, integration, and analysis capabilities for science and/or engineering data over a decades-long timeline; (2) continuously anticipate and adapt to changes in technologies and in user needs and expectations; (3) engage at the frontiers of computer and information science and cyberinfrastructure with research and development to drive the leading edge forward; and (4) serve as component elements of an interoperable data preservation and access network" (*34*).

NSF funded the first two DataNets in August of 2009. Both Data Conservancy and the Data Observation Network for Earth (DataONE) are highly collaborative teams representing many universities and disciplines. In the fall of 2011, NSF funded three additional projects – Sustainable Environment through Actionable Data, Terra Populus, and DataNet Federation Consortium (*35*).

As one of the first DataNets, DataONE serves as an exemplar for how a domain-specific research data organization can emerge and address the needs of is research community. The next sections of this chapter review the history, organization and activities of DataONE.

The Context for DataONE

In the 1980s, U.S. policy identified "Grand Challenges" as "fundamental problems in science and engineering, with broad applications, whose solutions would be enabled by high-performance computing resources…" Over the decades, the definition has grown to include the idea that solutions will require advances in computational models, data and visualization techniques, and collaborative organizations that unite disciplines (*33*). Among the Grand Challenges listed by NSF are prediction of climate change, water sustainability, and understanding biological systems (*33*). Studying these challenges demands the expertise of scientists from a wide range of domains, as well as experts from other disciplines, including social sciences. Verifiable and persistent data are essential to these researchers, a condition that increases the need for technology that supports data collection, quality control, data description, storage, integration, analysis, visualization, and preservation. This means there is a growing need for powerful tools that will enable all stakeholders -- scientists, academic researchers, government decision-makers, industry leaders, non-governmental leaders and the public – to engage the data.

People around the world are facing environmental, social, and technological challenges related to climate variability, altered land use, population shifts, and changes in resource availability (*36*). Environmental issues are complex and are studied by scholars in many different disciplines, resulting in the publication of results in different domains (*3, 37*). This separation makes data accessibility and sharing difficult. Scientists are working to understand these problems and to provide information that can be used to construct solutions. However, a major challenge in tackling global problems is finding a way to achieve a global perspective (*3*). Currently, there are high barriers of both time and cost for researchers who must retrieve content from multiple data repositories, in order to use that content in meta-analyses or for comparison with new studies, and then publish the output. Notably, the output may result in a journal article, or the publication of a dataset in a repository where others may similarly retrieve and utilize the data. This means that data may be viewed as both inputs and products of scientific research. What is needed is an infrastructure that provides scientists around the world with the proper tools to collaborate and analyze complex data, so they can concentrate on conducting science, rather than spending their time and energy devising processes to share and access the data.

DataONE is designed to address these needs. It is a multi-institutional, multinational, and interdisciplinary collaboration that serves the biological, ecological, and environmental research communities by building and supporting cyberinfrastructure that enables data intensive science. DataONE is structured to foster interaction across the many domains involved in environmental research by: (1) helping scientists to identify tools that will facilitate their workflow; (2) developing tools to assist scientists and other stakeholders; (3) building community across the various stakeholders; and (4) addressing the technical and sociocultural aspects of the scientific process.

While data access is the engine driving data-intensive science, addressing all data management issues requires looking at both technical and sociocultural solutions. Technical solutions focus on tools and technologies that enable data collection, storage, preservation, and access. Socio-cultural solutions must address improving the way people engage data (e.g. creation, use and re-use) including understanding their perceptions, attitudes, and behaviors. For example, scientists often do not know about the tools available to them for data management, metadata creation, and data preservation, and even if they do, they often do not use them (*38, 39*). Another major sociocultural issue is the lack of compatible data practices (*3*). Therefore, efficient tool development must consider the sociocultural issues, to be sure that the tools will have services that stakeholders want and can easily use.

As noted earlier in this chapter, trusted data can enable new science, and results from the analysis of the data can be used for evidence-based decision-making. DataONE's organizational structure supports the full data lifecycle by focusing on providing tools, training, and policy guidance to assure that data will be robust, persistent, and accessible. DataONE's goal is to "ensure the preservation of and access to multi-scale, multi-discipline, and multi-national science data," (*36*) while providing tools and best practices (*40*) to address data management challenges. These challenges include data loss (by natural disaster, format obsolescence, or orphaned data), scattered data sources, data deluge (the flood of increasingly heterogeneous data), poor data practices, and data longevity (*41*).

DataONE Vision and Mission

DataONE's vision is to be "...used by researchers, educators, and the public to better understand and conserve life on earth and the environment that sustains it" (*42*). The DataONE mission is to "Enable new science and knowledge creation through universal access to data about life on earth and the environment that sustains it" (*42*). This is being accomplished by concentrating on three core areas: (1) the provision of a toolkit for data discovery, analysis, visualization, and decision-making; (2) the provision of easy, secure, and persistent data storage; and (3) the facilitation of community engagement by scientists, data specialists, and policy makers (*42*).

The objectives for meeting this mission include:

(1) using the available cyberinfrastructure to provide coordinated access to current databases;
(2) creating a new global cyberinfrastructure that contains both biological and environmental data coming from different sources (research networks, environmental observatories, individual scientists, and citizen scientists);
(3) changing the culture of science by encouraging responsible cyberinfrastructure practices through education and training, engaging citizens in science, and building global communities of practice.

DataONE's Organizational Structure

In fall of 2007, when Dr. William Michener of the University of New Mexico assembled the team to answer the NSF DataNet call for proposals, he created a highly collaborative environment that engaged a wide range of disciplines and institutions. This has served as a basic tenet of the organization as it has matured. From a disciplinary perspective, environmental sciences are the central focus, with strong channels of interaction with other disciplines, including information and computer science.

Within each of these disciplines is a rich diversity of expertise. The environmental sciences include scientists from biology, ecology, environmental sciences, hydrology, and biodiversity. The information and computer science members include specialists in informatics, computer engineering, computer sciences, information sciences, information management, information technology, and library sciences. The organization is designed to expand to accommodate other disciplines that interface with environmental data. For example, sociologists may be interested in using the data to enrich their studies of migration and urbanization, or economists may be interested in data to augment their studies of natural resource allocation.

Institutional diversity is represented by the fact that active participants reflect eleven different kinds of institutions. These categories, along with some examples, are: (1) academic institutions, (2) research networks (e.g., the Long Term Ecological Research Network and the South African Environmental Observatory Network); (3) environmental observatories (e.g., the National Ecological Observatory Network, and the USA-National Phenology Network); (4) synthesis centers (e.g. National Center for Ecological Analysis and Synthesis, the National Evolutionary Syntheses Center, Oak Ridge National Laboratories, and the National Center for Supercomputing Applications); (5) government organizations (e.g., the U.S. Geological Survey, and NASA); (6) academic libraries; (7) international organizations (e.g., Global Biodiversity Information Facility); (8) data and metadata archives (e.g., ORNL Distributed Active Archive Center for Biogeochemical Dynamics, and the Knowledge Network for Biocomplexity); (9) professional societies (e.g., Ecological Society of America); (10) NGOs; and (11) the commercial sector (e.g., Amazon, Microsoft, IBM, and Intel) (*43*).

DataONE is a lean organization with a small managerial team based at the University of New Mexico to oversee the coordination of the organization from both the technical and sociocultural perspective. The managerial team is composed of the Principal Investigator, an executive director, a director of development and operations, and a director of community engagement and outreach. The director of development and operations oversees the work of a core team of designers and developers. There are also two advisory bodies – the External Advisory Board and the DataONE Users Group.

The External Advisory Board (EAB) is composed of approximately a dozen experts who are at the forefront of the environmental science, information science, and computer science communities. The EAB provides strategic direction and guidance to help assure that DataONE is in a position to address the challenges and potentials of the environmental science domain. The EAB helps to facilitate activities that increase community engagement and encourage sustainability.

The DataONE Users Group (DUG) meets annually to discuss the needs of the environmental science research community and the emerging technical challenges and opportunities that may be addressed through DataONE. This may involve DataONE products, tools, and services being used to advance education, research, and policy. The DUG is composed of members of the environmental science community, including Earth observation data creators and users. The DUG also includes members from the other stakeholder groups – libraries, data centers, government agencies, academic institutions, and research networks. The DUG held its first meeting in December 2010, and in the summer annually thereafter.

DataONE includes a large number of engaged volunteers whose work is coordinated through the working group model. Working Groups are discussed in the next section.

DataONE Working Groups

Working Groups are a key means for DataONE to engage in research and educational activities through the work of their members. Working groups serve as a means for bringing together a diverse array of experts to work intensively on targeted topics related to environmental researchers. This model assures that key issues are tackled from multiple perspectives, while allowing the evolution of research and educational activities. Each working group of ten to twelve members is composed of scientists, computer scientists, information scientists, library professionals, academic researchers, educators, and government and industry representatives. Each working group has developed a charter that states its purpose, outlines roles and responsibilities of its members, and identifies proposed work to be conducted and the timeline for its completion.

There are eleven working groups in Data ONE, and each is categorized as focusing on either cyberinfrastructure (federated security, preservation and metadata, integration and semantics, distributed storage, provenance and workflow) or community engagement (sustainability and governance, sociocultural issues, community engagement and education, public participation in science and research) issues. However, two working groups -- Usability & Assessment, and Exploration, Visualization, & Analysis – directly engage in both cyberinfrastructure and community engagement activities.

DataONE Data Lifecycle

The concept of a data lifecycle is useful because it identifies how data flow through the research process. DataONE is using the data lifecycle to focus on strategic decisions about developing tools, creating coordinated educational modules, and providing guidance for policy. There are many data lifecycle models in existence, each of which has value and strengths. DataONE participants reviewed many of these and took the approach of adopting a lifecycle model that places data as the focal point and illustrates the different stages that data may pass through, although data may skip a stage or stages (*43*). The people interacting with the data at each stage may vary, and it is unlikely that the same person will be working with the data at all stages.

There are eight stages of this data lifecycle, starting with *planning,* when a scientist outlines how he will conduct his research and collect data. This step has become a key activity for scientists seeking NSF funding, since NSF, in 2011, began requiring that a data management plan be submitted with each proposal. The second stage is *collecting*; the scientist collects data either in the field or laboratory. The next step is *assuring* the quality of the data. *Describing* is the fourth stage, as the data must then be described by metadata, so that the data may be more easily found and used in analysis. This is a point in the cycle that presents some issues, as scientists often use metadata schema that have been developed for their project, rather than employing the specific domain metadata standard. The fifth stage is about *preserving* the data, by depositing them into a trusted repository. Once in the repository, *discovering* the data becomes possible, as others may access the data. Data modelers or other scientists might then search for and access individual data, *integrating* multiple data sets and then *analyzing* them. In some instances, the original scientist who collected the data may continue using them, (skipping both the discovering and integrating stages).

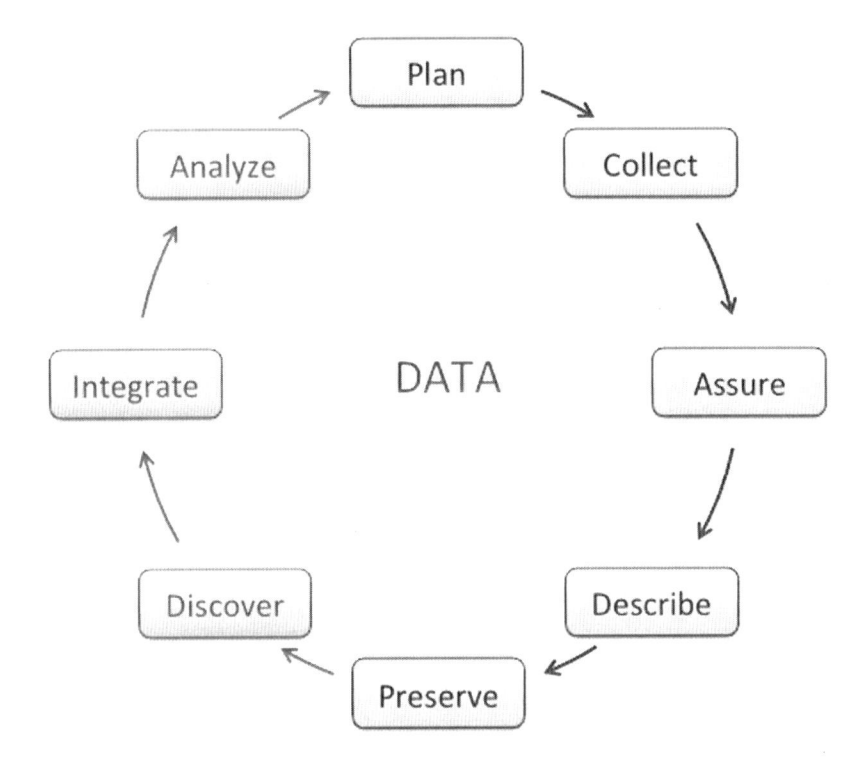

Figure 1. DataONE's data lifecycle illustrates how data moves through eight unique stages. Source: DataONE.

DataONE Cyberinfrastructure

DataONE was developed with three premises: (1) DataONE should support distributed management through a relationship between member nodes to enable replication, caching, and discovery of data across these repositories, supporting the goals of preservation, robustness, and performance. (2) DataONE software should provide benefits for scientists and data providers today and be adaptable to tomorrow's needs. (3) DataONE activities should support and use existing community software, emphasizing free and open source software.

DataONE's cyberinfrastructure is based on three core elements: coordinating nodes, member nodes and the investigator toolkit. Coordinating nodes enhance interoperability among the member nodes by providing network-wide services, such as indexing and replication. DataONE initially hosted three geographically dispersed coordinating nodes: one at the University of New Mexico, one at the University of California Santa Barbara, and one at the University of Tennessee in collaboration with the Oak Ridge National Laboratory. A selection of the kinds of utility services that coordinating nodes facilitate are: member node registration services, metadata indexing, coordinating and monitoring data replication, providing global user identity services, providing log aggregation services, and monitoring node and network health.

The member node network includes data centers, science networks, other organizations such as academic libraries, and government agencies. Member nodes can choose to expose their data to the DataONE network, and may also choose to provide computing resources or services to the broader DataONE community.

The investigator toolkit provides access to customized tools that scientists are already using, and to a range of tools that support scientists at each stage of the data lifecycle. These tools enable interaction with the DataONE infrastructure.

The DataONE cyberinfrastructure is designed to have the flexibility to address important challenges of a distributed network of nodes. The concept of connecting existing data and metadata repositories means that there must be a way to integrate the services that already exist and are tailored to the needs of each unique community. Currently, a researcher will often have separate accounts for each repository and must have knowledge of different tools to access and work with the data. In addition, repositories typically have different ways to identify data and metadata, which hinders data citation, duplicate identification, and provenance tracking.

Eight Challenges for Cyberinfrastructure

DataONE's cyberinfrastructure design specifically addresses eight major challenges (36). These challenges, and a descriptive overview of how they are being addressed, are listed below.

1. **Inconsistent service interface specifications:** A common set of interfaces is required to achieve interoperability in finding and retrieving data from the Member Nodes. A key aspect of the cyberinfrastructure was the design and implementation of consistent service interfaces, which are a fundamental requirement for technical interoperability between the DataONE infrastructure components.
2. **Lack of reliable unique identifier production and resolution:** Ensuring global uniqueness of identifiers, and assurance of global resolution of arbitrary kinds of identifiers, can be achieved through a variety of mechanisms. DataONE employs a simple, scalable mechanism that leverages the Member Node APIs to generate and maintain a catalog of identifiers in use, along with the nodes on which the content resides.
3. **Data longevity and availability is dependent on repository lifespan:** Despite the best of intentions, repository lifespans are often determined in conjunction with funding cycles or other financial/political factors. As a result, it is difficult to ensure that content will remain available to researchers in the future. Data longevity and persistent access is

important to the continuity of data use and reuse. The DataONE cyberinfrastructure addresses this problem by requesting that Member Nodes maintain copies of content from other Member Nodes where compatible data sharing agreements have been made.

4. **Inconsistent search semantics and effectiveness:** In the simplest sense, the goal of a search is to accurately identify a set of science metadata and data objects that precisely match a user's query. DataONE's initial implementation follows the approach of extracting a common set of attributes from science metadata and building a search index that is then exposed via APIs and web user interfaces. As the volume and breadth of data available through DataONE increases, search precision becomes increasingly critical. Therefore, DataONE is exploring how to incorporate semantic search, where the intent of the search term is taken into consideration along with the value. The efficacy of semantic search, however, is limited only by the richness of science metadata provided by Member Nodes and by the completeness of available ontologies.

5. **Varying service interactions and data models:** Researchers do not all use the same data model. The process for storing, modifying, and managing content varies between repositories, as do the conceptual models used to define associations between structures, such as metadata, data, annotations, packages, and collections. Such heterogeneity reflects the diversity of data collected from a range of experimental designs, devices, and purposes. However, there does appear to be a set of patterns exhibited by data structures commonly used in the Earth system sciences that can guide the generally expected behavior of data management systems, without negatively impacting the specialized behavior necessary to better serve specific communities. DataONE leverages the common service APIs supported by Member Nodes to promote consistent processes for routine data management operations.

6. **Access to quality metadata limits reuse of data:** The development of metadata tools, techniques, and principles is a significant focus of both technical and community engagement aspects of DataONE. The quality of science metadata directly affects the precision of search operations and the degree to which automated integration across data sets can be achieved. Data are expensive to capture and process so that they may be used in scientific analysis. Numerous metadata standards have been developed over time to address the difficulty of communicating specific information that can help researchers with the reuse of data. There is considerable benefit to ensuring ongoing reuse of existing data, both from an economic point of view and also because, as dataset accessibility continues, so does the opportunity for synthetic analyses and re-analysis. All content being added to the DataONE federation has "science metadata" associated with it. Additionally, an exact copy of the metadata is stored on the Coordinating Nodes, where it is both indexed to support search operations, and preserved to ensure longevity, without requiring that all Member Nodes must support the same metadata standard and syntax.

7. **The lack of shared identity and access control policies:** The goal is to protect Member Node data that reside within the system from intentional and non-intentional harm. This includes preventing unauthorized viewing of private data, alteration or deletion of another user's data within the system, and protecting services and other system resources from malicious activity that often occurs on the Internet.
8. **Difficulty in placing data near analysis, visualization, and other computational services:** Ideally, all data would be located close to the processing capabilities needed to perform the types of visualization, analysis, and processing required by researchers. However, this is not normally the case, and it is often necessary to move data to applications or vice versa, to perform the desired analyses. An important early outcome from the Exploration, Visualization & Analysis Working Group was the recommendation that DataONE infrastructure should interact closely with high performance computing facilities such as those implemented by the TeraGrid nodes.

Stakeholders and Community Engagement Activities

Stakeholder communities in DataONE are categorized into primary and secondary roles. Scientists are the primary stakeholders and they are a heterogeneous group. DataONE chooses to characterize scientists by how they "do" science rather than by domain in order to encourage the idea of practicing integrative science. Therefore scientists are placed in groups based on the environment in which they practice science. There are five *science research environments*: academia, government, private industry, non-profit and community. It is understood that scientists in private industry are likely to be restricted from sharing data because of proprietary concerns so they may not be able to interact with DataONE as data producers.

There are five groups of secondary stakeholders who interact in the data-intensive environmental science (Michener et al., 2011).

(1) Libraries and librarians are prioritized as the most important secondary community since they help scientists in all five science research environments negotiate data management issues. Libraries have a very broad definition including the full range of information-centric agencies and services.
(2) Administrators and policy makers influence science through funding programs and policy that may facilitate or hinder research. This group exists at the national, state and local level.
(3) Publishers and professional societies engage in activities that facilitate dissemination of research results and data.
(4) Think tanks develop evidence-based position papers or policy suggestions.

(5) The public includes citizen scientists, citizen activists, K-12 teachers, informal educators and curriculum builders. These are important stakeholders since they are the communication channel between science and the public.

Working groups who are categorized as part of the community engagement activities have already made contributions in reaching out to these stakeholders and helping DataONE successfully interact with these people. These activities are overseen by the DataONE executive team, particularly the Director of Community Engagement. The Community Engagement and Education Working Group has been instrumental in creating resources that identify best practices (*40*) for all aspects of data management and existing tools (*44*) that can assist researchers and data managers. This group has also been active in creating and hosting workshops for scientists at a variety of venues.

The Usability and Assessment Working Group has been active in surveying the various stakeholder groups to create baselines that establish current knowledge about data management and current data practices. Surveys have been conducted with scientists, librarians, libraries, government agencies, and data managers. Results are being widely shared with the community-at-large including both scientists and information professionals in libraries and data centers. This group has also been engaged in usability testing of DataONE products.

The Sociocultural Working Group has been working to address issues related to the culture of science. This group has helped define the data lifecycle, and the stakeholder community. It has also been active in creating personas to identify how different stakeholders might interact with DataONE. This group has also addressed communication strategies within the organization.

These examples of how the environmental science community is defined and being engaged serves as a portrait of how the cyberinfrastructure can be introduced to the research community so it is more likely to be adopted and used.

Collaborations and Synergies

DataONE recognizes that it is part of a larger data ecosystem where others are also working to address these issues of long-term data management, reuse, discovery, and integration. In order to maximize the effort of all of these projects, it is important to communicate and collaborate. DataONE is actively working and communicating with the other DataNet projects, and is also engaged in discussions with other projects that are targeting very specific technical topics. One strength that is aiding this collaboration is an overlap in participation between members of the various projects; both projects are sharing information about developments for data management and preservation, as well as sharing information about how these other projects might leverage the DataONE cyberinfrastructure.

DataONE has established many collaborations. DataONE has joined the Federation of Earth Science Information Partners as a Type I partner. DataONE is also exploring collaborative relationships with other projects, including: the Filtered-Push project (http://etaxonomy.org/mw/FilteredPush), the Scientific

Observations Network (SONet http://www.sonet.com/), Semantic Tools for Ecological Data Management (SemTools https://semtools.ecoinformatics.org/), TeraGrid (transitioning to XD/XSEDE), and the Avian Knowledge Network (AKN http://www.avianknowledge.net/content/).

There are also synergistic activities that DataONE is involved with, including the development of the DMPTool, which helps researchers create data plans online. The DMPTool (https://dmp.cdlib.org/) original partner institutions include four libraries in the United States and the United Kingdom's Digital Curation Centre.

In the future, DataONE envisions ever-strengthening collaborations involving more associated disciplines. For instance, possible areas for expansion include researchers studying migration and urbanization, such as sociologists, and natural resource allocation, such as economists. DataONE's goal, and challenge, is to create the cyberinfrastructure that can address multi-faceted environmental issues and mobilize the engagement of all of the interested parties.

Learning for the Future

Sound data is the foundation for good research and for results that can be used for evidence-based decision-making. Sound data begins with planning research activities to include good data practices throughout the data lifecycle, including data collection, quality assurance, and data description. Placing data in a repository is an important part of the lifecycle, but just being in a repository doesn't guarantee persistent access to these data, without careful planning and long-term support. Access is essential for the data to be found and used in any kind of analytical activity. Addressing the eight challenges associated with data requires a comprehensive, cohesive program, and nations around the world are building organizations to coordinate data management activities on a broad scale.

As one such initiative, DataONE, provides insights for successfully implementing initiatives of this type. While DataONE is focused on one scientific domain, environmental science, there are valuable lessons that can be applied to other large-scale data management initiatives:

1. Communication, both internal and external, is essential. Internal communication is needed to fully utilize the expertise of all those involved, especially if the initiative is highly distributed, a circumstance which is very likely for most data-centric projects. External communication is needed to link the project with other projects that are proximate in terms of domain, discipline, technical solutions, or sociocultural issues. Having strong communication channels facilitates learning about the developments in these proximate projects, and also provides a means to reach out to groups that could benefit from project activities. A carefully crafted communication plan can facilitate strategic objectives and ultimately may help foster success of the project in the data ecology.
2. It is important to utilize existing tools. Building on the established habits of data producers and users helps to facilitate the adoption of a new

system and new best practices. For example, scientists are currently using tools, which means that substantial development has already been accomplished and user communities are established. Loyalty to these existing tools and practices can be leveraged to more quickly build a critical mass of researchers who will engage with the new initiative.
3. Tools serve different parts of the data lifecycle, and there is a need to help data producers and users at each stage of the lifecycle. Specialized tools that are customized for the community's needs will allow data producers and users to select the best way to engage data at the point they enter the data lifecycle. The overall initiative can provide the context of the "one-stop shop" where this toolkit is housed.
4. The community should be consulted during development. A tool or an organization shouldn't be built in a vacuum, since it is unlikely to be successful if it doesn't meet the need of its community. Planning for communication allows community input during design and development and can keep projects on-track, although care should be taken to avoid scope creep. There is also a need to plan for encouraging community engagement with what is built. The idea of "if I build it, they will come" doesn't often work with products. Instead there must be a definite plan for bringing the community to the project and helping them to see how they will benefit from participating.
5. Watch for scope creep. It is easy for the project to expand beyond the original goals. Clearly define the goals and objectives, then regularly assess the activities to keep on-track.
6. Identify risks and potential mitigation strategies. Then, be sure to review them and use this review to inform development decisions.
7. There is hardware, software and wetware. Often the hardware and software components of a project get the most attention. Be sure to remember that people are part of the equation. The culture of the community involved in the project will dictate the best way to do this, so it is essential to get to know the community.
8. Leadership is important. Initiatives of all sizes, but particularly those that are large, require the creation of a shared vision and then the development of a small, committed, and effective team to help everyone move toward the realization of the vision.

References

1. Utility, C.o.E.t., Age, I.o.R.D.i.a.D., and Sciences, N.A.o. *Ensuring the Integrity, Accessibility, and Stewardship of Research Data in the Digital Age*; The National Academies Press: Washington, DC, 2009.
2. Hey, T., Tansley, S., Tole, K., Eds.; *The Fourth Paradigm: Data-intensive Scientific Discovery*; Microsoft Research: Redmond, WA, 2009.
3. Hunt, J. R.; Baldocchi, D. D.; van Ingen, C. Redefining Ecological Science Using Data. In *The Fourth Paradigm: Data-intensive Scientific Discovery*;

Hey, T., Tansley, S., Tole, K., Eds.; Microsoft Research: Redmond, WA, 2009; pp 21−26.
4. Lynch, C. Jim Gray's Fourth Paradigm and the Construction of the Scientific Record. In *The Fourth Paradigm: Data-intensive Scientific Discovery*; Hey, T., Tansley, S., Tole, K., Eds.; Microsoft Research, Redmond, WA, 2009; pp 177−183.
5. Core Techniques and Technologies for Advancing Big Data Science & Engineering (BIGDATA); Engineeering, D.f.C.I.S., Ed., National Science Foundation, Washington, DC, 2012.
6. Weiss, R.; Zgorski, L. J. Obama Administration Unveils "Big Data" Initiative: Announces $200 million in New R&D investments; Office of Science and Technology Policy, Washington, DC, 2012.
7. Service, A.N.D. Home, 2012.
8. Strategy, N.C.R.I. Providing eResearch infrastructure. In *An Australian Government Initiative*, 2010.
9. Strategy, N.C.R.I. eResearch infrastructure. In *An Australian Government Initiative*, 2010.
10. Strategy, N.C.R.I. The Super Science Initiatives. In *An Australian Government Initiative*, 2009.
11. Service, A N.D. Our Approach, 2012.
12. Forschungsgemeinschaft, D. From the Notgemeinschaft der Deutschen Wissenschaft to the Deutsche Forschungsgemeinschaft, 2012.
13. Winkler-Nees, S. Promoting Accessibility to Research in Germany: Funding Initiatives, Projects, and Perspectives. In *Research Data in Germany Informational Workshop*, Washington, DC, 2011.
14. Berlin Declaration. Max Planck Society, 2003.
15. Systems, C.o.S.L.S.a.I., and Management, S.o.I. Recommendations for Secure Storage and Availability of Digital Primary Research Data; Deutsche Forshungsgemeinschaft, 2009.
16. Pieters, J. The SURF Foundation. In *Ariadne*, 1996.
17. SURF. Access to research data, 2012.
18. Services, D.A.a.N. Home, 2012.
19. aparsenexchanges. DANS − Data Archiving and Networked Services of Royal Netherlands Academy of Arts and Sciences.
20. Beagrie, N.; Lavoie, B.; Woollard, M. DANS − Data Archiving and Networked Services. In *Keeping Research Data Safe*, Vol. 2; Charles Beagrie Limited: Salisbury, U.K., 2010.
21. Services, D.A.a.N. About, 2012.
22. Centre, D.C. Home, 2012.
23. Beagrie, N. A Continuing Access and Digital Preservation Strategy for the Joint Information Systems Committee (JISC) 2002−2005, 2002.
24. Centre, D.C. History of the DCC, 2012.
25. Committee, J.I.S. JISC in Transition, 2012.
26. Centre, D.C. DCC Phase 3, 2012.
27. Lyon, L.; Rusbridge, C.; Neilson, C.; Whyte, A. DCC SCARP: Disciplinary Approaches to Sharing, Curation, Reuse and Preservation. Iin *JISC Final Report*, Version 1.7; 2010.

28. Lippincott, J. K. Coalition for Networked Information (CNI). In *Encyclopedia of Library and Information Science*, 3rd ed.; Bates, M. J., Maack, M. N., Eds.; CRC Press: Boca Raton, FL, 2010.
29. Information, C.f.N. History, 2012.
30. Information, C.f.N. New Video: National Status of Data Management, 2012.
31. Foundation, N.S. About, 2012.
32. Council, N.S.F.C. Cyberinfrastructure Vision for 21st Century Discovery, National Science Foundation, 2007.
33. Cyberinfrastructure, N.S.F.A.C.f. The NSF-ACCI Task Force on Grand Challenges, National Science Foundation, 2011; p 88.
34. Sustainable Digital Data Preservation and Access Network Partners (DataNet), National Science Foundation, 2007.
35. Search Results (DataNet awards), National Science Foundation, 2012.
36. DataONE. Proposal, University of New Mexico, 2009.
37. Dozier, J.; Gail, W. B. The Emerging Science of Environmental Applications. In *The Fourth Paradigm: Data-intensive Scientific Discovery*; Hey, T., Tansley, S., Tole, K., Eds.; Microsoft Research: Redmond, WA, 2009; pp 13−19.
38. Tenopir, C.; Allard, S.; Douglass, K.; Aydinoglu, A. U.; Wu, L.; Read, E.; Manoff, M. Data Sharing by Scientists: Practices and Perceptions. *PLoS ONE* **2011**, *6*.
39. Kuipers, T.; van der Hoeven, J. Insight into Digital Preservation of Research Output in Europe; Insight, P., Ed.; 2009; p 83.
40. DataONE. Best Practices, DataONE, 2012.
41. Allard, S. DataONE: Facilitating eScience through collaboration. *J. eScience Librarianship* **2012**, *1*.
42. DataONE. What is DataONE? DataONE, 2012.
43. Michener, W. K.; Allard, S.; Budden, A.; Cook, R.; Douglass, K.; Frame, M.; Kelling, S.; Koskela, R.; Tenopir, C.; Vieglais, A. Participatory design of DataONE: Enabling cyberinfrastructure for the biological and environmental sciences. *Ecol. Informatics* **2011**, *11*, 5–15.
44. DataONE. Software Tools, 2012.

Chapter 5

Activities of Regional Consortia in Planning e-Science Continuing Education Programs for Librarians in New England

Donna Kafel[*]

e-Science Project Coordinator, Lamar Soutter Library,
University of Massachusetts Medical School,
Worcester, Massachusetts 01545
[*]E-mail: donna.kafel@umassmed.edu

In 2009, the libraries of the five University of Massachusetts campuses initiated a series of professional development programs to help New England science, health sciences, and engineering subject librarians build the knowledge and skills that are needed to support e-Science pursuits at their institutions. These programs have expanded to include the following annual events: an e-Science symposium, a Professional Development Day, and a Science Boot Camp. Alongside these conferences, the Lamar Soutter Library at the University of Massachusetts Medical School initiated a collaborative e-Science Portal for New England Librarians, partnered on a grant to develop frameworks for a data management curriculum, and established the Journal of eScience Librarianship. This chapter describes e-Science, its impact on libraries, and examines the e-science continuing education programs and research sponsored by a consortium of New England science, health sciences, and engineering librarians.

© 2012 American Chemical Society

Background of e-Science

Dramatic advances in digital technologies in the late 20th and early 21st centuries have created a ripple effect in the ways that scientists conduct research, collect and analyze data, and communicate their findings to colleagues and scholarly communities. Increasingly scientists utilize automated instrumentation like remote sensors, gene sequencers, simulation programs, micro-arrays, and computerized modeling in their research work. Large scale research investigations such as genomic sequencing and astronomical sky surveys have generated data sets of a magnitude and granularity exceeding any that could have been spawned by paper and analog photography (*1*). While the specific instrumentation may differ from one discipline to another, a key feature of all of these technologies is that they rapidly produce massive amounts of raw digital data. This accelerated generation of data in conjunction with the availability of the internet has enabled scientists to globally collaborate and quickly collect, analyze, and disseminate their findings—fast forwarding a scholarly process that used to take months or years. More than ever teams of scientists are working on a single project. Add to this scenario desktop access to journals and databases, and the result is today's research era of data-intensive scientific discovery known as the "Fourth Paradigm"— where "all of the science literature is online, all of the science data is online, and they interoperate with each other (*2*)." The research methodology intrinsic to the Fourth Paradigm is referred to as e-Science.

This new methodology of e-Science is data centric, data driven and collaborative. The term e-Science was first used in 1999 by John Taylor, Director General of Research Councils in the UK, to describe the "global collaboration in key areas of science and the next generation of infrastructure that will enable it." (*3*). Key features of e-Science include the adaptation of technologies for computation, modeling, sensing, data analysis, visualization, and collaboration in scientific research work. The availability of the internet and communication technologies has lowered geographical barriers and fostered virtual collaboration and team science. Through the internet, researchers can rapidly disseminate digital data sets. This in turn has fostered data sharing and an unprecedented level of access, promoted interdisciplinary teamwork on complex problems, and enabled other researchers to use data for different purposes than what the creator of the data had envisioned. Researchers and students reuse raw data to explore new or related hypotheses, often integrating the data with other datasets for analysis. The greater scientific community and the general public benefit from the sharing of data: it encourages multiple perspectives, enables scrutiny of findings, discourages fraud, aids in the training of new researchers, and increases the efficiency of funding by avoiding duplication of effort and resources. (*4*).

While the term e-Science often implies the computational work of large research teams, it is just as relevant for "small science", hypothesis driven research led by a single investigator or small research group that generates and analyzes its own data. (*5*).

Interestingly, the term "e-Science" has been more widely adapted by library and information science professionals than by the research community. Science researchers do not commonly describe the nature of their research endeavors as "e-Science". They may not use any terminology at all to describe the e-Science phenomenon. The use of computational tools and methodologies has become so ingrained in science that many researchers simply acknowledge it as the way that they "do science".

In the context of this chapter, the term e-Science includes all natural and physical sciences, health and other applied sciences, and technological disciplines.

Libraries and e-Science

As science becomes increasingly cognizant of data's potential for advancing research, all players in the traditional infrastructure of scientific research and communication have been strategizing potential roles in the science data landscape. National libraries, research funding agencies, universities and research libraries, software and publishing industries have all been exploring ways to address the data deluge (6). As historic leaders in the advancement of knowledge, universities have borne significant responsibility for the long-term preservation of knowledge through their libraries. Among the many participants in scholarly communication, librarians have been recognized for their expertise in organizing, enhancing and disseminating information, and have been identified as logical partners in the stewardship of digital data. A few university libraries have initiated shared data archives. According to Clifford Lynch, director of the Coalition of Networked Information (CNI), "these projects unite groups of people who usually don't work together: scientists and scholars on one side and library and IT folks on the other, are all feeling their way for the right roles for everybody." (7). The Association of Research Libraries (ARL) notes a need for new partnerships and collaborations among domain scientists, librarians, and data scientists in order to better manage digital data collections, and advocates expanded library roles in preservation and curation services of digital data sets.

Developing library infrastructures that support data curation and preservation activities requires a retooling of library services. Traditionally research libraries serve as custodians of "downstream knowledge"-- organizing and maintaining collections of post-research publications such as conference proceedings, journal articles, and books. Data science and management diverges from the text-oriented systems that still dominate library roles in science communication and publishing. With e-Science comes a new impetus to develop library infrastructures that support the "upstream knowledge" component of the research lifecycle early on in the scientific process. Reconfiguring library workflows so that librarians are involved in documentation during the earliest stages of research would help to ensure long-term preservation of data (8).

In Figure 1, Data and Publication Life Cycles, the steps of the pre-publication phase (i.e. upstream) and the publication phase (i.e. downstream) of the data and publications life cycle are illustrated. (6). While most librarians are well acquainted with the post-publication phase, for many the pre-publication phase of research during which data is created, collected, managed, and analyzed is unchartered territory.

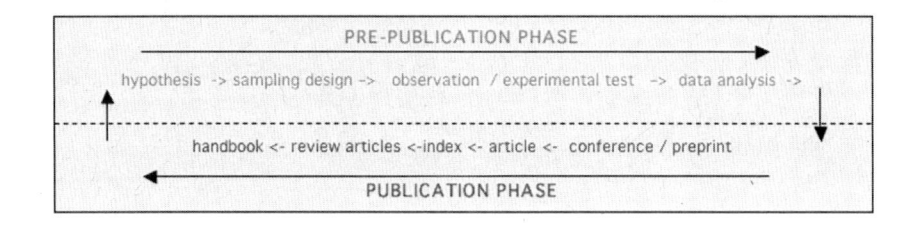

```
┌─────────────────────────────────────────────────────────────────┐
│                      PRE-PUBLICATION PHASE                    ▶ │
│  hypothesis -> sampling design -> observation / experimental test  -> data analysis -> │
│ ▲-------------------------------------------------------------- │
│        handbook <- review articles <-index <- article <- conference / preprint   ▼ │
│ ◀                                                               │
│                        PUBLICATION PHASE                        │
└─────────────────────────────────────────────────────────────────┘
```

Figure 1. Data and Publication Life Cycles. (Reproduced with permission from Anna Gold, ©2007, CNRI)(6).

In 2006, recognizing that libraries would need to adapt to changing conditions brought about by networked science, the Association of Research Libraries (ARL) appointed a task force to raise awareness and position research libraries to participate in e-Science. One of the task force's key findings is that librarians need to be actively engaged with their user communities more than ever before, and to do this, librarians need "to not only understand the concepts of a domain, they also need to understand the methodologies of scholarly exchange" (9).

It is important to note that many research libraries have already demonstrated an understanding of new methodologies of scholarly exchange by initiating outreach services through various ventures such as managing institutional repositories, publishing scientific journals, and collaborating with campus research and computing groups to develop and maximize the reach of research information networks. As digital collections and desktop accessibility become more prevalent, these ventures illustrate how librarians have successfully filled niches that address the information needs of their research communities. Although the format of information has changed, the call for librarians to collect, gather, organize, and make information accessible to those who need it still remains (10). Furthermore, just as academic librarians conduct information literacy classes to students, they can also address the gap in science data literacy through teaching data management best practices and by providing consultation services to their research community.

While there have been multiple reports that advocate library involvement in e-Science, many science, health sciences, and engineering librarians find themselves at a loss as to exactly how they might participate in e-Science and what skills they can offer. How can a subject librarian help a scientist with a data management plan? What advice does a librarian give to a research group on the management of multiple versions of intricate data sets? How does a librarian become knowledgeable about metadata standards for different science domains? Where does a librarian find models for implementing working relationships with researchers? There is a need for professional development and working models that address the everyday tasks of e-Science librarianship. Purdue librarians Garritano and Carlson (11) identified skill sets that librarians new to e-Science should expect to adapt or develop:

- Library and Information Science expertise
- Subject expertise
- Partnerships and outreach (both internal and external)
- Participating in sponsored research
- Balancing workload

In ARL's 2007 report "Agenda for Developing e-Science in Research Libraries", the ARL's Joint Task Force on Library Support for E-Science identified desired outcomes for positioning the research library community as partners in the development of e-Science. One addresses criteria for a future research library workforce: "Knowledgeable and skilled research library professionals with capacity to contribute to e-science and to shape new roles and models of service". The task force suggested the following strategy: "build a library workforce with relevant new skills and knowledge about emergent forms of documentation and research dissemination." One proposed action for implementing this strategy is to "pursue science librarian skills to meet the needs of e-science." (*12*).

In October 2008, the ARL and the Coalition for Networked Information (CNI) hosted the "Reinventing Science Librarianship" forum. During this forum, several speakers shared their ideas on what the science librarian in the near future would look like in terms of skills, capacities, and institutional positioning. One speaker, Rick Luce, Vice Provost and Director of University Libraries, Emory University described his vision of a future science library in which multi-skilled information management teams could be created "on the fly", and embedded librarians would collaborate with research teams or departments to provide timely and holistic advice on documentation throughout the research process. Luce commented that emerging forms of scientific practice will require different kinds of library support at different times. One point of consensus among all the forum presenters was that "the fundamental role of the science librarian needs to expand to incorporate skills related to organizing and manipulating data and data sets." (*13*).

With an interest in exploring new library roles in e-Science and heeding the recommendations from the recent ARL report and forum, in 2008 a small group of library administrators and science librarians from the five University of Massachusetts libraries met. At this meeting the group discussed ways that the UMASS libraries could develop e-Science library services and engage in the state's Life Science Initiative (*14*). During the group's early discussions, the need for affordable continuing education opportunities for science, health sciences, and engineering librarians was raised. Acting on this need, the group laid the groundwork for a series of events that has significantly raised New England librarians' level of understanding of science subjects and e-science; and promoted intercampus working relationships. The history of this early group, its strategy for promoting e-Science librarianship in the New England region, and its findings are featured in the next section. The remainder of the chapter will include details on the components of the University of Massachusetts and New England Area continuing education and research program in e-Science.

Promoting e-Science Librarianship in New England

While New England research libraries were contemplating ARL's agenda for initiating library engagement in e-Science, in June 2008 Massachusetts Governor Deval Patrick enacted the Life Sciences Initiative, a one billion dollar investment package to enrich and strengthen the state's globally recognized leadership in the life sciences. The strategy of the Life Sciences Initiative was "to bring together industry, academic research hospitals and public and private colleges and universities to coordinate this effort, spur new research, strengthen investments, create new jobs and produce new therapies for a better quality of life." (*15*).

Highlighted as a partner for driving further innovation in life science research, the University of Massachusetts (UMASS) established a Life Sciences Task Force (LSTF), which was charged with "crafting a university-wide aspirant vision in the life sciences and promoting inter-campus collaboration." (*16*).

The greater academic community, the academic health sciences libraries and the community of ARL science libraries within the Boston Library Consortium responded positively to this call for inter-campus collaboration. In the fall of 2008, library directors and nine science librarians from the five UMASS campuses met to explore how the individual campuses could collaborate to be included in the Massachusetts LSTF's future funding allocations. This group, which has since come to be known as the "UMASS 5", brainstormed ways that they could expand their library services to support the networked research efforts of the Life Sciences Initiative. Each member of the UMASS 5 shared a common concern as to how her library would meet ARL's outcome of building a library workforce with an e-Science capacity. Many ideas and concerns were raised, including the need to educate librarians in research areas relevant to the Life Sciences Initiatives.

Following the initial meeting of the UMASS 5, in October 2008, the National Networks of Libraries of Medicine for the New England Region (NN/LM NER) held a meeting with sixteen New England Resource Library Directors. When asked about their interest in learning more about e-Science and exploring opportunities for inter-campus collaboration, all directors expressed a need to learn more about how their libraries could be positioned to participate in the scientific research arena. They agreed to attend and invite their library staff to attend a regional symposium on e-Science and instructed the Lamar Soutter Library at the University of Massachusetts Medical School (UMMS), to take the lead in planning this e-science symposium. This consensus by the New England Resource Library Directors cemented the UMASS 5's and the NN/LM NER's primary strategy to foster librarian education and collaboration in e-Science in the New England region. Funding was obtained from the NN/LM NER and the Boston Library Consortium for this daylong symposium, which would serve as a resource workshop and think-tank for regional academic health sciences and science librarians (*17*). The Lamar Soutter Library then proceeded to plan the first annual e-Science symposium, which was eventually held in April 2009.

While the planning activities of the UMASS 5 group initially focused on the data aspects of e-Science, several librarians in the group noted that they had no formal training in the sciences. The group conceded that this lack of a science background was a common one hindering many librarians from approaching researchers and engaging in the scientific arena. This led to an overall acknowledgment that having disciplinary knowledge is a fundamental first step for e-Science related library services. Not being versed in the terminology of a science or understanding its key concepts are obstacles for librarians attempting to assist researchers in using databases or subject-specific tools. Noting the financial and time constraints that preclude working librarians pursuing a second degree in a science, the UMASS 5 group discussed the need for affordable continuing education in the sciences for librarians and came up with the idea of hosting a Science Boot Camp for librarians. (*18*).

With a directive and financial support from their library administrators and the NN/LM NER, and the Boston Library Consortium, the UMASS 5 group has since organized four Science Boot Camps and assisted with programs organized by UMMS and the NN/LM NER. These programs offer affordable professional development opportunities designed to educate librarians in the sciences, promote readiness for e-Science engagement, and provide opportunities for librarians to engage and collaborate with each other and scientists. In all, the 2009 inaugural events that paved the way for an ongoing series of New England e-Science learning opportunities include:

- e-Science symposium on April 6
- Stem Cell Workshop Professional Development Day on May 13
- Science Boot Camp on June 24-26, 2009

These first three educational events have promoted working relationships among campus libraries in the New England region and spurred the development of a regional e-Science Librarian Community of Interest (COI). Collaboration within this e-Science COI has paved the way for an e-Science portal for New England Librarians, a broader engagement of New England librarians beyond the original UMASS 5 in planning yearly Science Boot Camps, a partnership between the University of Massachusetts Medical School and Worcester Polytechnic Institute on an IMLS National Leadership Planning Grant to develop frameworks for a data management curriculum and identify user needs for a shared data repository, and the *Journal of e-Science Librarianship*. Each of these initiatives will be discussed in this chapter.

Strategy for New England e-Science Library Programs

The preceding section presents the chronology of the planning events that sparked ideas for the e-Science symposium, Professional Development Day, and Science Boot Camp. The primary strategy for supporting these programs was to promote librarian education and collaboration in the area of e-Science. Over the past four years as the New England region became more active and aware of e-Science issues, this primary strategy diverged into six components: Common identity, Roadmap, Tools, Continuing Education, Dissemination of findings, and Scholarship & Research. Table 1 illustrates these strategy components and their activities.

Table 1. e-Science Strategies and Activities of the New England Consortia

e-Science Strategy	Activities
Community Engagement (common identity)	Community of Interest (COI)
Roadmap for Libraries (how to engage in e-Science)	e-Science Symposium
Tools	e-Science Portal for New England Librarians, Frameworks for a Data Management Curriculum
Continuing Education	Professional Development Days, Science Boot Camps
Dissemination of Findings and Intellectual Examples	Journal of eScience Librarianship
Scholarship & Research	Assessment of competencies, survey of educational programs in data curation and management

Each of these six components addresses specific facets for promoting and supporting e-Science librarianship in the New England region. Building an e-Science Community of Interest (COI) is an ongoing activity that began with the first meeting of the UMASS 5 and has grown to include all the attendees of the e-Science symposia, Science Boot Camps, and Professional Development Days. A goal for developing this COI is to build a community of librarians with a common understanding of e-Science library roles in the region. Currently the e-Science Community of Interest includes 155 New England science, health sciences, and technology librarians.

The Roadmap for Libraries strategy aims to increase awareness amongst New England librarians of the importance of e-Science and how libraries can support scientific research. At each of the e-Science symposia, presenters have discussed examples of library engagement in supporting science—in various ways from NSF "big science" grants to an example of a library research data working group that is scanning the needs of science researchers across campus in order to plan relevant library research support services. The poster presentations at the e-Science symposia have allowed individual librarians to disseminate news of their e-Science related library projects in a setting conducive to one-on-one discussions—fulfilling the roadmap strategy and helping to build relationships among the e-Science COI.

Tools provide resources specifically targeted for librarians that support their engagement in e-Science services. One such tool is the e-Science Portal for New England Librarians. The portal is a centralized online website that provides links to information on e-Science, sciences, data curation, data management planning, courses, workshops and opportunities. The Frameworks for a Data Management Curriculum, with its lesson plans and research cases, serve as a tool for librarians and science, health science and technology faculty for implementing formalized data management instruction to their students.

Continuing education supports working librarians by bringing them up to speed on emerging fields and new research methodologies. Planning for these continuing education activities are heavily based on attendee evaluations at previous events—resulting in a series of professional development days and Science Boot Camps that feature topics regional librarians want to explore and understand better.

The Journal of eScience Librarianship provides a forum for dissemination of scholarly findings and intellectual examples. The journal's mission is to advance the theory and practice of librarianship with a special focus on services related to data-driven research in the physical, biological, and medical sciences. It welcomes original articles related to outreach, collaborative, and educational aspects of e-Science librarianship from librarians around the globe.

The sixth and final strategic component is Scholarship and Research. Work in this area is underway. The UMMS e-Science team has been assessing the region's e-Science COI in order to identify requisite data management competencies that area librarians need to engage in e-Science. The team has also conducted a survey of library school programs that provide data curation and management training.

The following sections of the chapter will further describe the events and tools created by the New England consortia.

e-Science Symposium for New England Librarians

Since 2009, the Lamar Soutter Library at UMMS, the NN/LM NER and the Boston Library Consortium have co-sponsored four annual e-Science symposia for New England Librarians. Each symposium is a daylong conference featuring presentations by nationally recognized experts and local librarians. The ultimate goal for the e-Science symposium initiative has been to develop a strategy for a regional collaboration for delivery of e-Science resources and services.

The inaugural e-Science symposium held in 2009 was intended to increase awareness among New England area librarians of the importance of e-Science and the role of the library in supporting scientific research. Eighty science, health sciences, and technology librarians from thirty-eight regional libraries attended. The morning's introductory panel featured two health sciences librarians and one researcher. Their presentations included defining e-Science, its major issues, and the roles librarians can play to support e-Science at their institutions. This panel was followed by presentations from a bioinformatics librarian at a major biomedical research institution and a researcher from the University of Massachusetts Medical School International Stem Cell Registry, which demonstrated two specific applications of e-Science to symposium attendees. A subsequent presentation about the University of Massachusetts Center for Clinical and Translational Sciences explained Clinical and Translational Science and its goal to strengthen collaboration among all sciences programs and encourage interdisciplinary collaborations that will enable and accelerate translation of research discoveries in labs into clinical practice. The ability of high-computing technologies and interdisciplinary networking between labs and departments to extend research work beyond the confines of a single institution was emphasized, along with the impetus for librarians to keep abreast of research projects within their institutions so that they can identify potential areas for library involvement.

During this first e-Science symposium, there was a pivotal afternoon breakout discussion session that prompted attendees to begin thinking about e-Science issues and how their libraries could build working relationships with their research communities. One key area of discussion during this breakout session addressed the credentials, knowledge and competencies that librarians need to engage in e-Science. Attendees noted that these would include specific technical skills for data stewardship, disciplinary content knowledge, and collaboration skills. This topic evolved into further conversation on what types of continuing education events and resources would help practicing librarians acquire necessary competencies and knowledge. Feedback from the discussion groups showed a strong preference for a combination of in person classes and online resources. The idea for a centralized online e-Science web portal that would provide links to resources, tutorials, and tools and include a discussion forum for librarians working in diverse science research institutions in New England was widely embraced by the group. Taking these responses into consideration, in May 2009 the Lamar Soutter Library began planning the development of the e-Science portal for New England Librarians, which will be described further on in the chapter.

Since this first e-Science symposium in 2009, the programs for successive symposia have featured both nationally recognized and local speakers—librarians, computer scientists, life and physical scientists presenting on topics ranging from large-scale National Science Foundation data projects, library involvement in planning campus data services, institutional approaches to data curation, linking data to publications, data repositories, science data literacy, and current research projects that illustrate data creation and management in lab settings. Beginning in 2010, the symposia have also featured poster presentations by mainly regional librarians (over the years this has broadened with more librarians outside the NE region participating) discussing their respective e-Science projects. To view the programs and presentations for the 2011 and 2012 e-Science symposia, go to the University of Massachusetts and New England Area Librarian e-Science symposium website at http://escholarship.umassmed.edu/escience_symposium/. Programs and slides presentations from the 2010 and 2009 symposia can be accessed through the University of Massachusetts and New England Area Librarian e-Science Initiatives webpage at http://library.umassmed.edu/escience_initiatives.

Next Steps

Topics for future e-Science symposia will include library roles in high-performance computing initiatives, qualities of effective library collaborations, and characteristics of library leadership in successful collaborations.

Professional Development Days

Professional development days provide area librarians with an affordable continuing education opportunity to learn about a specific area of scientific study or aspect of science librarianship. The following topics have been covered in these professional development day workshops for New England Science Librarians:

- Exploring Stem Cell Research: What does it mean for Librarians?
- Nanotechnology in the Health and Applied Sciences: Implications for Librarians and Researchers
- Scientific Data Management
- Metadata Day

The Exploring Stem Cell Research professional development day was held in May 2009 and was hosted by the Lamar Soutter Library and the Center for Stem Cell Biology and Regenerative Medicine at UMMS. The program consisted of an introduction to stem cell biology, an overview of the International Stem Cell Registry http://www.umassmed.edu/iscr/index.aspx that resides at UMMS, a discussion on the intellectual property and patent issues related to stem cell research, bioethical considerations; a tour of the Stem Cell Center, and an afternoon table discussion of opportunities for librarians in the sciences.

The following year in May 2010, the University of Massachusetts Amherst Libraries sponsored the Nanotechnology in the Health and Applied Sciences professional development day. The focus of this program was nanotechnology in health and other applied sciences. Librarians learned about nanotechnology terminology, tools, and information resources used by nanotechnology researchers, such as *InterNano* (http://www.internano.org/) an information portal managed by one of the science librarians in the UMASS 5 group.

Scientific Data Management was the topic for the 2011 Professional Development Day. Jian Qin, director of the e-Science program at the Syracuse iSchool delivered a comprehensive presentation that included background history of scientific approaches that has lead up to the current data-centric e-Science methodology, descriptions of datasets, a survey of science metadata standards, conducting data interviews, and recommendations on how librarians can build working skills in e-Science. Her presentation can be viewed on the Scientific Data Management subject guide at http://libraryguides.umassmed.edu/content.php?pid=176769&sid=1496225.

The fourth in this series of professional development days, Metadata Day, has yet to be held as of the writing of this chapter. The idea for devoting a program to metadata came from attendees at the 2011 Scientific Data Management workshop, who wanted to explore the role of metadata in more depth. The program for this day will feature an overview of metadata and its role in enabling discoverability, access, and interoperability of information, and afternoon breakout sessions at which local librarians from diverse science research libraries will present their metadata-related work projects.

The subject guide for Metadata Day can be viewed at http://libraryguides.umassmed.edu/content.php?pid=319888.

Next Steps

The proposed topic for the next professional development day is an exploration of tools that support research collaboration, such as research information networks.

Science Boot Camp for Librarians

Initiated in 2009, the goal of Science Boot Camp for Librarians has been to provide more in-depth, affordable science education and networking opportunities for librarians over a two and a half day period. Each boot camp features three sessions in which local science faculty share their expertise by providing an overview of a science followed by presentations of relevant research projects within the field. Faculty presenters gear their lectures toward non-specialists. This has provided boot campers with an understanding of the concepts and terminologies of science disciplines, which will ultimately enable them to better engage with their research faculty.

Science Boot Camp's casual atmosphere fosters dialogue between faculty and librarians. This has been an all-around positive experience for both faculty and librarians--enabling librarians to directly ask faculty questions about their area of expertise and get a better understanding of their research processes--and informs faculty about the role of libraries in supporting scientific research. Each boot camp has been held at a different university and has featured three science sessions. At the request of attendees at the first boot camp, a capstone lecture was added for the last day of boot camp. The capstone sessions highlight innovative projects of librarians relevant to e-Science. Links to the Science Boot Camp library guides and recorded presentations from the Science Boot Camps (except for the 2009 Science Boot Camp recording) are posted on the Science Boot Camp page of the e-Science Portal for New England Librarians http://esciencelibrary.umassmed.edu/science_bootcamp.

Table 2 provides a summary of the host campuses, featured science topics, and Capstone topics for the boot camps as of June 2012.

Table 2. Summary of Science Boot Camp Topics 2009-2012

Year	Host	Science Topics	Capstone Presentation
2009	University of Massachusetts Dartmouth	• Bioinformatics • Geographic Information Systems • Nanotechnology	(Had not been initiated yet)
2010	University of Massachusetts Lowell	• Genetics • Remote Sensing • Climate Change	• DataStaR: Cornell's data repository • e-Science portal
2011	Worcester Polytechnic Institute	• Robotics • Astronomy • Epidemiology	• Science informatics at the Marine Biological Lab at Woods Hole Oceanographic Institute Library
2012	Tufts University	• Neuroscience • Organic chemistry • Data visualization	• ARL e-science Institute • Library school programs

Planning Process

Having conceived the idea for Science Boot Camp, the UMASS 5 science librarians planned the first two boot camps in 2009 and 2010. As noted in Table 2, these boot camps were hosted at the UMASS campuses in Dartmouth and Lowell. The planning process for each boot camp gets underway about ten months prior to the actual boot camp and involves three face to face meetings at UMASS Medical School in late summer, fall and winter, followed by phone conferences in the spring months leading up to boot camp. Except for the inaugural 2009 boot camp, planning starts with a review of attendees' evaluations from the most recent boot camp. In these evaluations, attendees provide feedback on the venue, activities, science and capstone sessions, along with recommendations for changes and topics for future boot camps. The group makes note of the attendee feedback when selecting the science topics and planning arrangements for the next boot camp.

A concerted effort is made to host Science Boot Camp at a different campus each year; enabling each librarian in the group an opportunity to host a boot camp on her campus. Selection of the venue for boot camp is done as early as possible in order to secure conference facilities at the host campus and set the date. Traditionally Science Boot Camp is held in early June when students have left campus for summer break so that overnight boot campers can stay in student dormitories. Efforts are also made to plan boot camp dates that do not conflict with national library association conferences. The librarian serving as Boot Camp host coordinates the accommodations, food, meeting rooms, A/V recordings of presentations, and check in process at her campus.

Once the group selects three science topics for the boot camp science sessions, planning group members suggest faculty subject experts from their respective campuses to present. Following this meeting, they contact these faculty members to inform them about Science Boot Camp and gauge their interest and availability. At the next meeting, the group discusses faculty responses and collectively decides which specific faculty members to invite to present. The agenda for the capstone and ideas for librarian presenters are also planned. Over the course of late fall and winter, invitations are issued and the agenda is planned in more detail. Tasks and roles are assumed by specific librarians in the planning group. These include publicizing boot camp, setting up a boot camp subject guide, setting up and processing registrations, planning social events for boot camp, and developing evaluation forms and science session merit badges. During Science Boot camp, planners help the host with last minute details such as setting up folders and displays, meeting, greeting, assisting attendees and presenters, maintaining attendee lists, facilitating social events, and collecting evaluation forms.

After the 2010 Science Boot Camp, the UMASS 5 noted that expanding its boot camp planning group to include librarians outside of UMASS would provide new opportunities for collaboration, fresh ideas, and open up boot camp presentations to faculty from other universities. The group invited attendees working in other New England institutions to participate in planning future Science Boot Camps via a Science Boot Camp follow-up online survey sent to attendees. Included in the survey was a section where interested librarians could fill out their names, institutions, and contact information. Eight librarians indicated interest in participating in boot camp planning in their survey responses. Reviewing these responses, the UMASS 5 group realized that it needed to set criteria for librarian involvement in Science Boot Camp in order to ensure that librarian participants would have institutional support and commitment. The following criteria were established:

For a librarian to participate in the Science Boot Camp planning group, the following criteria need to be met:

- The librarian must be currently employed at a library.
- This library or its parent institution must provide co-sponsorship funding for Science Boot Camp (*19*).

Having secured financial sponsorship from their parent institutions, librarians from Worcester Polytechnic Institute, Tufts University, University of Connecticut, and College of the Holy Cross have joined the planning group. With this expansion, the UMASS 5 Science Boot Camp planning group has evolved into the "Science Boot Camp Planning Group."

Funding for Science Boot Camp is made possible through co-sponsoring libraries, the NN/LM NER and the Boston Library Consortium. The combined efforts of the librarians who have collaborated to plan Science Boot Camp have been crucial to its success. Sharing a passion for science, they are committed to providing affordable quality continuing education to the greater library community. The process of sponsoring boot camp demands a cohesive team effort to plan a budget, venue, science sessions, capstone, and social events; invite speakers, publicize, process registrations, and serve as hosts during the actual event. Each year an average of sixty librarians attend Science Boot Camp.

The librarians in the boot camp planning group have made a concerted effort to disseminate news of Science Boot camp through publications in library journals and presentations at various library conferences, both nationally and internationally (22).As word of boot camp has spread, it has sparked a keen interest among librarians both inside and outside the New England region to use the boot camp model for continuing education events in their communities of interest. One example of this is a Social Science boot camp that librarians at Tufts University initiated in 2011 (20).

Next Steps

Planning for the 2013 Science boot camp will commence in fall 2012. Decisions for future science topics for each science session will be based on campers' suggestions from the 2012 Science Boot Camp evaluations. With continued funding from the NN/LM NER, the Boston Library Consortium, and institutional sponsors, the Science Boot Camp Planning Group's long range plan is to continue the ongoing series of annual Science Boot Camps while maintaining affordable registration fees, and to offer sponsoring libraries opportunities to serve as Science Boot Camp hosts on their campuses.

e-Science Portal for New England Librarians

The e-Science portal for New England Librarians http://esciencelibrary.umassmed.edu/ is an openly accessible web resource with links to reports, white papers, articles, tutorials, conference presentations, and science primers that are relevant to e-Science librarianship. A component of the portal is the e-Science Community blog, where opinion pieces, reviews, and thoughts by contributing science librarians, and news, events and opportunities are posted. Any postings on the blog automatically feed to the portal's Twitter account @NERescience.

The content of the portal is divided among these main focus areas:

- **About e-Science**: e-Science Overviews, Researchers and Data, Cyberinfrastructure, Policy
- **E-Science and Libraries**: ARL reports, Library roles, Librarian Education, Assorted library blogs
- **Data Support Services**: Data management Planning, Data Repositories, Data Curation, Science Data Literacy, Research Information Networks
- **e-Science Community**: e-Science Community blog, New England Projects, National Projects, Organizations
- **Science Primers**: Links to overview tutorials in life and physical sciences; Science Boot Camp, Research Tools and Methodologies
- **About the Portal**: Scope statement, content selection criteria, information about the portal staff, advisory and editorial board members

As mentioned earlier in the chapter, attendees at the first e-Science symposium in April 2009 noted the need for a web portal to provide e-Science continuing education. In May 2009, the Lamar Soutter Library at UMMS obtained funding from the NN/LM NER to develop the e-Science portal. After a portal web team was established, one of the first steps in planning the portal was to assess what New England science, health sciences, and engineering librarians' e-Science learning needs were. This information was critical to guide the construction and scope of the portal. There were three key objectives for this needs assessment:

- To establish the need for an e-Science portal
- Examine what e-Science and data services librarians and libraries were currently providing
- Identify the educational background of the region's science, health sciences and technology librarians along with their educational needs and social media preferences in order to develop the portal scope and transmission mechanism

The assessment was developed as an online survey that was sent out to 168 unique libraries and individual medical, health sciences, and science and technology librarians who served or whose institutions served medical or interdisciplinary biomedical researcher patrons. Seventy-eight librarians responded to the survey. Overall, the responses to the survey revealed that a small number of New England libraries were currently engaged in e-Science activities at their institution or with other institutions, with a larger number of librarians seeing potential for future e-Science projects. These results indicated that the New England library community needed an e-Science web portal. Results also showed a regional demand for e-Science and data services tools and scientific content tutorials. Respondents indicated that they were comfortable with using a variety of educational web 2.0 tools for self-guided learning and online discussions (*21*).

Another early step in planning was to populate an advisory board of New England librarians that would define the scope of the portal, its major subject areas, aid in the selection of content editors, and guide the portal's overall development. A science librarian with a strong interest in e-Science and who managed a nanotechnology portal accepted an invitation to serve as chairperson for the advisory board. Invitations were then issued to library directors and librarians who advocated the development of e-Science library services in the region. Eight individuals from the following research institutions joined the advisory board: MIT, Worcester Polytechnic Institute, Massachusetts General Hospital, UMASS Medical School, Genzyme, and Yale University. Ex-officio members of the advisory board included the PI for the portal, portal coordinator, and the associate director of the NN/LM NER. The first meeting of the advisory board was held in October 2009. During this meeting the results of the portal needs assessment were presented, and the board discussed what the scope, audience, and topic areas would be for the portal. The advisory board created this scope statement:

The portal is designed for librarians working in research organizations that generate, share, store and/or use data for basic scientific research in the health, biological, and physical sciences. Bringing together resources on education, outreach and collaboration, current practices and e-science news—the portal provides librarians with the tools, knowledge and skills to effectively participate in networked science (22).

Two months after this initial meeting, the advisory board reconvened to plan charges for a portal editorial board and recommend experienced science subject librarians who could oversee the selection of content for the portal focus areas and participate in joint meetings with the advisory board to guide the portal web team in developing the portal. Upon the advisory board's recommendations, the portal coordinator sent invitations to the suggested subject librarian candidates and within a few months the editorial board was established. Members of the editorial board include science, health sciences and engineering subject librarians from Tufts, MIT, Northeastern, and UMASS Amherst, NN/LM NER, and a high school science teacher with an MSLIS who is currently pursuing a PhD in education. In April 2010, the editorial board and the advisory board met for their first joint meeting. During this meeting, the portal web team and the boards planned the delegation of portal content areas to specific editorial board members and established the process by which content would be selected, annotated, evaluated and posted on the portal. The portal web manager joined this meeting and presented his ideas on how he would initiate planning the structure of the portal with Drupal, an open source content management system.

The delegation of portal focus areas to the content editors is based on their area of expertise and interest. Two editorial board members submit content relevant to news, events and current projects. Two board members assumed responsibility for the development of the portal blog. The established workflow between editors and the portal coordinator includes the following steps:

1. Content editor reviews, selects and annotates resources (articles, papers, video tutorials, news items).
2. She then submits the content to the portal coordinator. This has been done mainly by e-mail. For the news/events content editor, it became much easier for her to post content on a social bookmarking site that the portal coordinator could regularly access.
3. The portal coordinator reviews the content to ensure it meets the content selection guidelines and falls within the scope of the portal. If she has questions about this, she discusses it with the content editors or an advisory board member. Once the coordinator approves the content, she forwards it to the portal site manager to post it.

After almost two years of planning, the portal was officially launched at the e-Science symposium in April 2011. During the afternoon session of the symposium, the project coordinator presented the portal and explained its key focus areas to symposium attendees. Since the launch, the portal staff reviews monthly usage statistics for both the portal and the e-Science community blog. Usage of the portal has increased steadily and globally since the portal was first launched. Usability studies have been conducted annually. Subjects of these usability studies include Simmons College GSLIS students enrolled in a science and technology resources course, and librarians working in New England research libraries. The results of these usability tests have informed the staff of ways they can revise the portal to enable users to find content and navigate the portal more efficiently.

The portal serves as a great working model of regional librarian collaboration. The synergistic expertise of administrative and subject librarians has facilitated the identification of critical learning needs and selection of quality content that will help librarians develop necessary knowledge, skills and competencies to engage in e-Science.

Next Steps

The idea of expanding the portal advisory and editorial boards to include librarians outside of New England was discussed at a recent board meeting. It was decided that for the near future, oversight of the portal will continue by a collaboration of New England librarians. Limiting the board members to New England librarians facilitates the hosting of affordable in-person biannual joint board meetings.

Currently the e-Science portal for Librarians is targeted specifically for librarians. This may continue, or a decision may be made to broaden the portal content to target scientific researchers as well as librarians. This idea will be open for discussion at a future joint meeting of the portal staff, advisory and editorial boards.

IMLS National Leadership Planning Grant: Planning Frameworks for a Data Management Curriculum and User Needs for a Student Data Repository

In August 2010, the Lamar Soutter Library at UMMS and the George C. Gordon Library at Worcester Polytechnic Institute were awarded a collaborative one year Institute of Museum and Library Services (IMLS) National Leadership Planning Grant and funding from the NN/LM NER to develop frameworks for a data management curriculum for undergraduate and graduate students in the sciences, health sciences, and technology disciplines; and to identify user needs for a collaborative data repository for data generated from student research projects. Reports about this grant, along with presentations and the project outputs are accessible via the project's website at http://library.umassmed.edu/imls_grant (*23*).

The impetus for this grant arose from discussions between the two schools' library directors who had compared notes on the lack of consistent data management standards in the research arena in general, and a local need for formalized data management instruction for students in the sciences. Both were well aware of frustrations voiced by faculty at their schools over data management issues in the lab and clinical settings. They also recognized the potential value data from student projects has for re-use in future projects. In addition to planning a data management curriculum, the two schools also wanted to explore available open source repository software systems to assess their usability for possible development of a shared repository for data generated by student projects.

Grant personnel included a Steering Committee made up of the two library directors who served as co-PIs, the project coordinator, an associate library director from UMMS, and the director of research computing at WPI; and an Education Committee of faculty and librarians from the two schools, and outside consultants (curriculum consultant, evaluation consultant, and an instructional design consultant).

Planning Frameworks for a Data Management Curriculum

The first phase in planning the curriculum was an extensive literature search of existing online data management curricula. Findings from this revealed very few curricula that were consistently accessible and targeted for students in the sciences (outside of library and information sciences).

In the second phase, project consultants conducted 50 interviews with students (30 freshman at WPI, 10 UMMS students, and 10 WPI graduate students) about data sharing and their current data management practices. The idea of sharing data was alien to many students, particularly freshmen undergrads. Two medical students at UMMS who had previously worked in labs for drug development companies stated that the concept of sharing data with other researchers outside of their organization was akin to disclosing trade secrets. One graduate nursing student described how she kept all data from subject interviews for a public health research project in files on her laptop, which she kept at her side at all times. Once the project was done, she destroyed these files. She was not aware of the option of stripping the data of personal identifiers so that it could be shared with other researchers without jeopardizing subject confidentiality. It was found that students maintain data in a variety of formats such as Excel, Sims 3, Word, Power Point, Adobe Illustrator, SPSS, and SAS. They often store data in their e-mails, in the cloud (e.g. Google Docs), on local drives, laptops, network drives, or external drives. There was a lack of standard naming conventions for directories and files, unless specific instructions were given by the PI or research supervisor. A common scenario in several research labs is the use of a paper lab notebook for specific types of data, as well as digital files. This juxtaposition of analog and digital data stored in separate locations can present challenges and potential gaps in integrating project data. The interviews with students from the two schools revealed many areas for potential data mismanagement, reflecting many findings from the literature search.

After the literature search and responses to student interviews were analyzed and the National Science Foundation's general requirement for data management plans was reviewed, the following learning objectives for the data management curriculum were identified:

By participating fully in this curriculum, the student should be able to:

1. Explain the need for managing/sharing research data, relevant public policies, and the lifecycle continuum for managing and preserving research data
2. Identify potential re-users, the value of your research data for re-use, and a dissemination strategy
3. Use an abbreviated data management plan or data curation profile to manage your research project data and define roles/responsibilities of research staff
4. Explain the range of research data types, stages, formats, and relevant software that may need to be managed and pre- served in your future research efforts
5. Identify what descriptive data needs to be documented in a standard way via metadata to allow your research data sets to be managed and preserved
6. Plan how to handle issues involved in securely storing research data in central databases, archives and/or repositories, backing it up, and managing access to your data
7. Explain legal (ownership) and ethical considerations related to data-sharing
8. Plan for issues related to long-term preservation, discovery, and re-use

These learning objectives were then translated into a plan for seven discrete course modules as noted in Figure 2.

The modular format of the curriculum is designed to be flexible in use for students at various educational levels (undergraduate, Master's/PhD). For example, an undergraduate working on his first research project may be required to review all seven course modules while a graduate student working in a new lab may be required to review modules 1-3. The curriculum is designed to be delivered in a variety of methods: video, online self-paced, and classroom instruction. These options allow faculty to customize course content so that it can be integrated into a range of learning environments.

Two UMMS faculty members on the project Education Committee suggested the addition of research case studies that depicted real life scenarios in lab, field and clinical research settings to the curriculum. Noting that work practices, terminologies, and data vary considerably from one discipline to another, they recommended to the Steering Committee that actual cases from a range of science, health sciences and engineering research areas be included in the curriculum. This would tie abstract data management concepts to real life situations that students could envision. The project Evaluation Consultant and a librarian from the Education Committee met with faculty members at both schools to elicit details of data management issues from their research experiences. From these interviews, the two wrote cases that illustrated data management issues in the following research settings: clinical behavioral health, biomedical lab research, orthopedic medicine, and an aerospace engineering lab.

Course Modules

Module 1
Overview of Research Data Management

Module 2
Data: Types, Stages, and Formats

Module 3
Metadata

Module 4
Data Storage, Backup, and Security

Module 5
Legal and Ethical Considerations

Module 6
Data Sharing and Reuse Policies

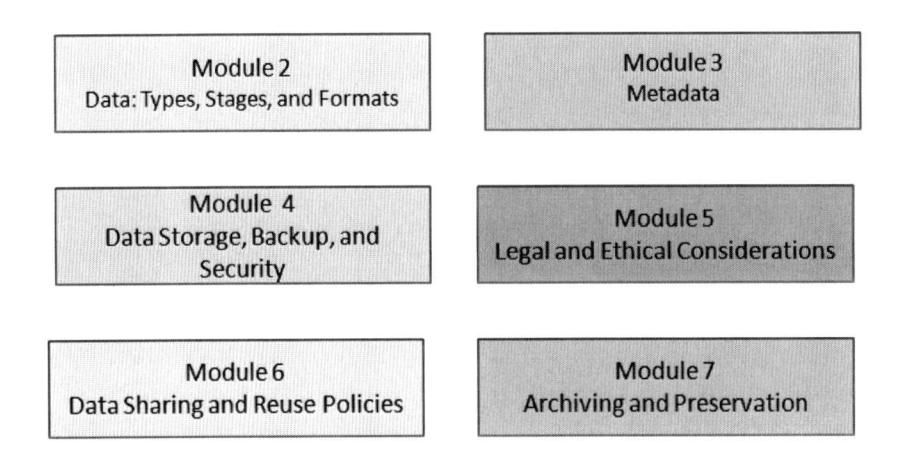

Module 7
Archiving and Preservation

Figure 2. Course Modules for Data Management Curriculum.

Lesson plans for the course modules were developed by librarians on the Education Committee and the Evaluation Consultant. Each lesson plan included specific learning objectives, lecture content, readings, activities, and an assessment. Content for Course Module 5 "Legal and Ethical Considerations for Research Data" was completely developed and integrated into a prototype online module featuring video and text instruction. This was done as a proof of concept for presentation to faculty at the two schools.

Identifying User Requirements for a Collaborative Data Repository

The Steering Committee investigated open source repository software systems that could be used to develop a data repository through a search of open access data repositories. They accessed data repositories to examine their user interface, analyze search functions, navigability, ease of access to data sets, and details about their software components. Findings were that many repositories were built on highly customized software, some with proprietary software, and of the open source software systems, DSpace, Fedora, and Islandora were the most prevalent.

WPI's Research Computing Services department conducted studies on DSpace, Fedora, and Islandora data repository software by loading data sets onto these three systems and evaluating the systems' user interface, analyzing search functions, tools, administrative requirements, available technical support and cost. Details of this testing can be viewed in the document "Evaluating User Requirements for Data Repository Software" at http://library.umassmed.edu/eval_user_reqs_data_software.pdf. The criteria and findings from the testing are outlined in "Matrix of User Requirements for Repository Software at http://library.umassmed.edu/user_reqs_matrix.pdf.

Next Steps

As of the writing of this chapter, the Lamar Soutter Library at UMMS has requested further funding for full implementation of the course content and pilot instruction of the modules at partnering institutions. These partnering institutions include UMASS Amherst, Tufts University, Northeastern University, and the Marine Biological Laboratory at Woods Hole Oceanographic Institute.

Journal of eScience Librarianship

In February 2012, the Lamar Soutter Library at UMMS launched the *Journal of eScience Librarianship, (JESLIB)*, an open access, peer-reviewed online journal whose goal is to advance the theory and practice of librarianship with a special focus on services related to data-driven research in the physical, biological, and medical sciences. JESLIB aims to promote the development of e-Science librarianship as a discipline and provide a forum for librarian discussion on issues related to managing, curating, preserving and retrieving clinical and science data. Original research papers, case studies, editorials, and conference proceedings from the annual University of Massachusetts and New England Area Librarian e-Science Symposium are featured in the journal.

The journal is an outgrowth of the series of e-Science conferences and outreach projects that have taken place in New England and outlined in this chapter. Like the other initiatives, *JESLIB* is a collaborative effort. The editorial board of the journal is made up of a team of librarians from the Lamar Soutter Library and an editorial consultant at the University of California Davis. Librarians engaging in e-Science services or projects in New England and other US regions serve as peer reviewers. (*24*).

The Journal of eScience Librarianship can be accessed at http://escholarship.umassmed.edu/jeslib. Since the launch of the first issue of JESLIB that featured the proceedings of the 2011 librarian e-Science symposium, JESLIB has received several manuscript submissions for its upcoming issue. These submissions have included original research papers, case studies, and reviews of e-Science workshops by librarian authors working in diverse research institutions across the US.

Next Steps

JESLIB was launched a few months prior to the writing of this chapter. Since then, its editorial board of UMMS librarians has been evaluating the processes that it implemented for publishing the first issue. The board has clarified editorial roles and responsibilities, fine-tuned the peer review process, and established editing and dissemination policies and procedures in order to build an efficient and successful working model for publishing future issues. Submissions for the next issue of *JESLIB*, which will be published in late summer of 2012 are currently being peer reviewed.

Scholarship and Research

The Lamar Soutter Library is just embarking on this strategic component of the e-Science program. In 2011, the library's e-Science team surveyed the NE region's e-Science librarian COI to assess the competencies needed by health sciences, science and technology librarians to engage in data curation and management and support e-Science research endeavors. Findings from the assessment revealed twenty requisite competency areas. Of these competency areas, the one with the greatest need for librarian training was the digital description and curation of large data sets *(25)*. In the six months between Fall 2011 and Spring 2012, the library's e-Science team conducted a survey of current data curation and management courses available in American Library Association-accredited Library and Information Science Programs in North America. Results of this survey have revealed a significant gap in educational programs in data management and curation. *(26)*.

Discussion

The e-Science symposia, professional development days, Science Boot Camp, e-Science Portal for New England Librarians, Frameworks for a Data Management Curriculum and User Requirements for a Collaborative Repository project, the Journal of eScience Librarianship, and e-Science scholarship and research are components of a long range strategic plan for e-Science learning and practice in New England. This strategy began with the first e-Science Symposium and has gradually developed over the last four years. The New England e-Science regional program now serves as a model of collaboration for other library consortia. Libraries interested in initiating similar ventures would benefit from understanding the key components that have contributed to the success of the New England regional e-Science program:

- **Clear purpose**: to promote and support e-Science librarianship through educational programs, development of an e-Science community of interest, and promotion of regional library partnerships in e-Science projects.
- **A lead institution with dedicated staff**: the Lamar Soutter Library at the University of Massachusetts Medical School.
- **Funding mechanism**: funding has been provided by the National Network of Libraries of Medicine for the New England Region, whose mission is to bring librarians together to support their continuing education needs. The Boston Library Consortium and The Institute

of Museum and Library Services have also funded e-Science program components.
- **Identified need and commitment from regional library directors**: contribution of resources (staff, funding, materials).
- **Regional interest**: Science, health sciences, and technology librarians in New England have identified e-Science librarianship as a common area of interest and are demonstrating a commitment to the e-Science program through ongoing attendance and participation.

A crucial first step in laying the groundwork for the e-Science program was securing a commitment to e-Science programming from the resource library directors of the Boston Library Consortium. Over time, this commitment has included financial sponsorship and donation of staff time and resources to the program. To ensure broad attendance at the first e-Science symposium, the Lamar Soutter Library sought and obtained commitment from New England Resource Library directors that each director would delegate two of their staff subject librarians to attend the event.

The e-Science program team reviewed the attendee evaluations for each of the inaugural events in 2009 (e-Science symposium, stem cell professional development day, Science Boot Camp). These reviews revealed a keen interest among regional librarians for further e-Science related educational programs and resources. Moreover they enabled the program team to identify regional librarians who were interested in helping plan and organize further events. The team followed up on these evaluations and issued invitations to these librarians to participate in the e-Science portal project as advisory and editorial board members. Since accepting these roles, the portal board members have shared their time and expertise by guiding the development of the portal, contributing content, promoting the portal, conducting usability studies of the portal, and advising the portal staff on next steps. Through their contributions to the portal project, the portal boards have demonstrated an extraordinary level of commitment to promoting e-Science librarianship in the New England region.

The original UMASS 5 Group and the Science Boot Camp Planning Group that evolved from it have also demonstrated this extraordinary level of commitment. To date the group has organized four very successful Science Boot Camps and will soon begin planning the fifth. The success of Boot Camp is reflected by the positive feedback from enthusiastic Science Boot Campers, the high percentage of campers who return each year, a surge in first time Science Boot Campers from diverse regions of the US and Canada, and use of the Science Boot Camp as a model for developing the Social Sciences Librarian Boot Camp at Tufts University.

The e-Science program has promoted awareness of e-Science, potential roles for librarians in providing e-Science related services at their institutions, and the competencies needed to engage in these roles, to librarians in the New England region. E-Science services can include data management instruction to students and faculty, consultation on data management plans for funding agencies, managing data collections in institutional repositories, participation in the development of research information networks, and data curation and preservation. Through its research and scholarship initiatives, the e-Science program has identified e-Science services currently underway and plans for future e-Science services at New England research libraries. Findings of the data management competencies assessment serve to inform future directions for the program's continuing education initiatives. Librarians who use the e-Science portal, the Journal of eScience Librarianship, and attend the e-Science symposia, professional development days, and Science Boot Camp can acquire knowledge and competencies that will enable them to effectively engage with researchers at their institutions and provide data support services. In addition to providing learning opportunities, the e-Science program has promoted collaboration among science, health sciences, and engineering librarians in the region. Successful collaborations in which library partners are able to leverage their resources and expertise will spur innovations in e-Science library services across and beyond the New England region.

Since the data management plan mandate for National Science Foundation grants was implemented in 2011, New England librarians have been asked to assist researchers in writing data management plans and providing data management instruction. In response to this increasing demand for library based data support services, there have been changes in librarian job descriptions and new positions such as data librarians have been created. Several New England libraries have partnered with their research and IT departments to plan campus-wide research data management services. Librarians involved in these endeavors and librarians who are preparing for future e-Science roles have come to rely on the resources provided by New England's e-Science program.

Conclusion

Technologies and the resulting outgrowth of digital data continue to evolve and expand in scale. Librarians serving the scientific community face both opportunities and challenges as they look for ways to tackle the data deluge and forge ahead in the nascent discipline of e-Science librarianship. Participating in research data management services requires a retooling of established library roles, workflows and competencies-- and a collaborative spirit. This collaborative spirit is alive and well in New England, where the combined efforts of a regional consortia of science, health sciences, and technology librarians have resulted in a repertoire of initiatives that will support e-Science professional development in New England and around the world.

Acknowledgments

Many thanks to the members of the initial UMass Five Campus Science Librarians group, the library directors of the UMASS 5 campuses, the Science Boot Camp planning group, and the librarians involved in the e-Science portal for New England Librarians project. Thanks especially to Elaine Martin, Sally Gore, Raquel Abad, Rebecca Reznik-Zellen, and Maxine Schmidt for their input. Components of the New England e-Science programs for librarians are possible due to the sponsorship from the National Network of Libraries of Medicine New England Region, the Boston Library Consortium, the Institute of Museum and Library Services, and the following New England institutions: the five University of Massachusetts campuses, Worcester Polytechnic Institute, University of Connecticut, Tufts University, and College of the Holy Cross.

The e-Science program for librarians in the New England region is funded by the National Library of Medicine, National Institutes of Health, Department of Health and Human Services, under Contract No. N01-LM-6-3508 with the University of Massachusetts Medical School. The Frameworks for a Data Management Curriculum and Requirements for a Collaborative Repository project is made possible by a grant from the Institute of Museum and Library Services and with funds from the National Library of Medicine under Contract No. N01- LM-6-3508. The Boston Library Consortium has cosponsored the e-Science symposia, professional development day workshops, and Science Boot Camps.

References

1. Association of Research Libraries. *To Stand the Test of Time: Long Term Stewardship of Digital Data Sets in Science and Engineering*, 2006, 15. http://www.arl.org/pp/access/nsfworkshop.shtml (accessed March 28, 2012).
2. Tenopir, C; Allard, S; Douglass, K; Aydinoglu, AU; Wu, L; et al. Data sharing by scientists: Practices and perceptions. *PLoS ONE* **2011**, *6* (6), e21101DOI:10.1371/journal.pone.0021101.
3. Thorpe, A. Environmental e-Science. *Phil. Trans. R. Soc.* **2009**, *367* (1890), 801−802. http://rsta.royalsocietypublishing.org/content/367/1890/801.full.
4. Rhoten, D. Dawn of networked science. *Chron. Higher Ed.* **2007**, *54* (2), B12.
5. Cragin, M. *Small Sciences Could Benefit from Better Data Sharing Practices*, 2010. http://www.lis.illinois.edu/articles/2010/09/small-sciences-could-benefit-better-data-sharing-practices.
6. Gold, A. Cyberinfrastructure, Data and Libraries. Part 1. *D-Lib Magazine*, 2007, Vol. 13, No. 9/10. http://www.dlib.org/dlib/september07/gold/09gold-pt1.html (accessed March 28, 2012).
7. Carlson, S. Lost in a sea of science data. *Chron. Higher Ed.* **2006**, *52* (42), A35. http://chronicle.com/article/Lost-in-a-Sea-of-Science-Data/9136 (accessed April 23, 2012).
8. Gold, A. Cyberinfrastructure, Data and Libraries. Part 2. *D-Lib Magazine*, 2007, Vol. 13, No. 9/10. http://www.dlib.org/dlib/september07/gold/09gold-pt2.html (accessed June 22, 2012).
9. Association of Research Libraries. *Agenda for Developing e-Science in Research Libraries*, 2007. http://www.arl.org (accessed March 28, 2012).
10. Gore, S. *Shaping Up: Boot Camp and Other Programs Addressing Professional Development Needs of Science Librarians*, 2011, 3. University of Massachusetts Medical School, Library Publications and Presentations,

Paper 123. http://escholarship.umassmed.edu/lib_articles/123 (accessed April 9, 2012).
11. Garritano, J. R.; Carlson, J. R. A Subject Librarian's Guide to Collaborating on e-Science Projects. *Issues in Science and Technology Librarianship*, 2009, 57. http://www.istl.org/09-spring/refereed2.html (accessed March 28, 2012).
12. Association of Research Libraries. *Agenda for Developing e-Science in Research Libraries*, 2007, 16. http://www.arl.org (accessed March 28, 2012).
13. Jones, E. Reinventing Science Librarianship: Themes from the ARL-CNI Forum, *Research Library Issues: A Bimonthly Report from ARL, CNI, and SPARC*, 2009, No. 262:13. http://www.arl.org/resources/pubs/rli/ (accessed April 23, 2012).
14. Abad, R. *New England Region E-Science Position Paper: Where We Have Been and Where We Are Going*, 2011, unpublished.
15. Massachusetts Life Science Center. http://www.masslifesciences.com/mission.html (accessed February 24, 2012).
16. University of Massachusetts Life Science Task Force. *A University-Wide Plan To Strengthen the Life Sciences and Promote Inter-Campus Collaboration over the Next Five Years.* http://media.umassp.edu/massedu/rsch/UMass_LSTF_Report_9-26-08.pdf (accessed February 24, 2012).
17. Rivera, R. *The University of Massachusetts and New England e-Science Symposium*, May 2009. Report to the National Networks of Libraries of Medicine New England Region (NN/LM NER), unpublished.
18. Schmidt, M.; Reznik-Zellen, R. *Science Boot Camp for Librarians: CPD on a Shoestring*, 2011, 2. http://conference.ifla.org/past/ifla77/200-schmidt-en.pdf (accessed April 26, 2012).
19. UMASS 5. *The UMASS 5 Science Boot Camp for Librarians*, unpublished.
20. Social Sciences Librarians Boot Camp. *The UMASS 5 Science Boot Camp for Librarians.* http://sites.tufts.edu/sslbc2012/ (accessed June 26, 2012).
21. Creamer, A.; Morales, M.; Crespo, J.; Kafel, D.; Martin, E. R. *Assessment of Health Sciences and Science and Technology Librarian e-Science Educational Needs to Develop an e-Science Web Portal for Librarians*, 2011. University of Massachusetts Medical School, Library Publications and Presentations, Paper 120. http://escholarship.umassmed.edu/lib_articles/120 (accessed April 20, 2012).
22. Portal Scope Statement, 2011. *e-Science portal for New England Librarians.* http://esciencelibrary.umassmed.edu/about (accessed April 20, 2012).
23. *Planning a Data Management Curriculum and Requirements for a Collaborative Repository*, 2012. http://library.umassmed.edu/imls_grant (accessed April 22, 2012).
24. Abad, R.; Gore, S. A.; Kafel, D.; Martin, E. R.; Palmer, L.; Piorun, M. E. *So You Want to Be a Publisher: Planning and Publishing the Journal of eScience Librarianship.* Poster presented at the University of Massachusetts and New England Area Librarian e-Science Symposium, April 4, 2012. http://escholarship.umassmed.edu/escience_symposium/2012/posters/1/ (accessed June 22, 2012).

25. Creamer, A.; Morales, M..; Crespo, J.; Kafel, D.; Martin, E. R. An assessment of needed competencies to promote the data curation and management librarianship of health sciences and science and technology librarians in New England. *J. eScience Libr.* **2012**, *1* (1), Article 4. http://escholarship.umassmed.edu/jeslib/vol1/iss1/4.
26. Creamer, A.; Morales, M.; Crespo, J.; Kafel, D.; Martin, E. R. A survey of LIS programs' scientific data curation and management courses. *J. eScience Libr.* **2012**, in press. http://escholarship.umassmed.edu/jeslib/.

Chapter 6

Interdisciplinary Data Science Education

Jeffrey Stanton,[*,1] Carole L. Palmer,[2] Catherine Blake,[2] and Suzie Allard[3]

[1]School of Information Studies, Syracuse University, Syracuse, New York 13244
[2]Graduate School of Library and Information Science, University of Illinois, Champaign, Illinois 61820
[3]School of Information, University of Tennessee, Knoxville, Tennessee 37996
*E-mail: jmstanto@syr.edu

Data scientists are information professionals who contribute to the collection, cleaning, transformation, analysis, visualization, and curation of large, heterogeneous data sets. Although some conceptions of data science focus primarily on analytical methods, data scientists must also have a deep understanding of how project data were collected, preprocessed and transformed. These processes strongly influence the analytical methods that can be applied, and more importantly how the results of those methods should be interpreted. In the present chapter we provide background information on educational challenges for data scientists and report on the results of a workshop where experts from the information field brainstormed on the educational dimensions of data science. Results of the workshop showed that data scientists must possess a breadth of expertise across three areas – curation, analytics, and cyber-infrastructure – with deep knowledge in at least one of these areas. Workshop participants also underscored the importance of domain knowledge to the success of the data science role. Additionally, the workshop highlighted a factor that differentiates data science from other professional specialties: the emphasis on serving the data needs of information users and decision makers.

© 2012 American Chemical Society

"Wanted, Data Scientist: Expected to work as a software developer and quantitative researcher, by driving the collection of new data, the refinement of existing data, and the analysis, interpretation, and communication of results to key team members." The preceding text paraphrases a recent job advertisement from a well-known Silicon Valley company. The job title itself, as well as the language in the advertisement exemplifies the emergence of a new hybrid area of applied practice focusing on the collection, cleaning, transformation, analysis, visualization, and curation of large, heterogeneous data sets. The widespread availability of varied, inexpensive (relative to historic norms) data, and the resulting proliferation of very large, collaboratively managed datasets has increased the need for professionals who can solve large scale information management problems for a range of users including scientists, engineers, policy makers, and business owners. The industry areas recruiting data scientists are highly diverse. Across these varied settings, a data scientist must understand how analytical methods fit within the entire data lifecycle, including data generation and preservation activities. Thus the tasks and problems of data science appear to require talents beyond a single discipline such as statistics.

For better or worse, however, the very term "Data Scientist" does seem to evoke images of statisticians in white lab coats staring fixedly at blinking computer screens filled with scrolling numbers, but this seems an unlikely scenario. First of all, statisticians do not generally wear lab coats: this fashion statement is reserved for chemists, biologists, and others who have to keep their clothes clean in environments filled with unusual fluids. Second, although professionals in every sector collect plenty of numeric data, a considerable amount of other information in science, engineering, education, government, and industry does not arrive as neat rows and columns of numbers and is not easily treated to statistical tests. Think of a hundred project folders full of paper forms, photographs, sketches, formulas, and handwritten notes or a hundred thousand PDF files containing reports with tables, graphics, and narratives: lots of data but little for a statistician to work with in these scenarios. In addition, data science covers the entire information lifecycle and requires a combination of technical and interpersonal skills necessary to understand existing information behaviors that surround data generation, access and reuse. Data scientists must have the skills to help users transform domain problems into questions answerable with existing data, to translate user data needs into technical specifications, to create and manage metadata, to envision and create appropriate data transformations and linkages, to manage data repositories, to work with data representation standards, to understand which analytical methods are appropriate given the existing data, and to know when to collect new data (*1–4*). Jim Gray, among the the first advocates of data intensive science, described the need to "support the whole research cycle – from data capture and data curation to data analysis and data visualization" (*5*).

This chapter presents the results of background research and a workshop session conducted on the educational requirements for data scientists. The background includes a brief account of the historical context that surrounds the emergence of the data science specialization as well as information from position postings showing some of the job characteristics of available data science positions. The workshop session was a brainstorming activity held at the February 2012 iConference – the annual conference of U.S. and international schools of information. The brainstorming activity brought together a substantial group of subject matter experts who developed consensus opinions concerning educational requirements in four areas of data science: domain knowledge, analytics, curation, and infrastructure.

Background

In June 2011, the McKinsey Global Institute released a comprehensive report entitled, "Big data: The next frontier for innovation, competition, and productivity" (6). The report detailed some of the current job market conditions for data professionals and projected existing trends in employment to understand future demands. The report claimed, "There will be a shortage of talent necessary for organizations to take advantage of big data. By 2018, the United States alone could face a shortage of 140,000 to 190,000 people with deep analytical skills as well as 1.5 million managers and analysts with the know-how to use the analysis of big data to make effective decisions." The McKinsey perspective reflects a view of data science that springs from industrial and corporate contexts, but nonetheless appears to represent a commonly held belief that demand for the profession is rising steeply.

Widespread recognition of the need for data science took approximately a decade to emerge. In 2001, John Taylor, the Director General of Research Councils at the Office of Science and Technology in Great Britain, articulated a vision for large-scale scientific collaboration that would be enabled by collaborative management of large datasets (7). Shortly thereafter, in the U.S., a National Science Foundation panel, headed by University of Michigan School of Information dean Dan Atkins, described similar sentiments, but expanded the scope beyond science and into engineering and industrial research and development through a newer term, "cyberinfrastructure", which refers to infrastructure based upon distributed computer, information and communication technologies (8). In both cases, the vision included recognition that a "data deluge" would serve as the driving force for investment in talent as well as technology.

The initiatives of the science establishment in the U.S. and U.K. and the McKinsey report represent endpoints over a broad swath of contested ground, particularly with respect to terminology and definitions. While the present chapter cannot hope to reconcile all (or perhaps any) of these differences, by focusing our attention on skills and occupational opportunities, we hope to sidestep some of these debates. Thus, while cyberinfrastructure, eScience, and data science are all labels that have been applied in overlapping spheres to a variety of interrelated subject matter, we here use data science simply as an umbrella term to direct attention to the confluence of challenges related to the management, analysis, and curation of large, heterogeneous datasets in a variety of contexts including science, engineering, education, government, and industry.

Example Applications

Certain industries, such as the insurance industry, have collected, maintained, and analyzed large data sets since before the computer age. Other industries, such as the airlines, have co-evolved with the development of contemporary information technology and so have put their data to work in increasingly sophisticated ways as the capabilities of the technology have grown. Within science and engineering, the use of sensor systems together with information technology has served as a primary driver of collection and curation of large data sets. The examples below typify the responsibilities of a data scientist who must consider the entire data lifecycle.

The data lifecycle begins with data collection. Although data collection varies across sectors, there are commonalities that make different application areas more similar than one might first believe. For example, in science and engineering, much data collection occurs through the use of instruments and sensors (e.g., seismic sensors in geophysics). Yet consider that industrial applications have led the way in the use of sensors in the form of barcode readers, radio frequency identification tags, and global positioning systems. In each case, an automated or semi-automated system provides an episodic or continuous stream of data based on the behavior of some object or objects of interest. A data scientist does not have to deploy the sensors herself, but she does have to know how the data were collected. The capabilities of contextualizing, cleaning, managing, and analyzing data all depend upon having the appropriate formats, linkages, and metadata. Raw data typically requires multiple transformations before data reuse is possible. In data mining applications, the difficulty and time involved in making these transformations may leave, "a gap between the potential value of analytics and the actual value achieved" (*9*). When these concerns are neglected early in the data collection process, the later jobs of working with the data become considerably more complex and expensive.

The data lifecycle continues with data management, preparation, organization, and distribution tasks that make the data ready for analysis and consumption by end users. This part of the lifecycle has been of particular interest in the scientific community, where big data problems in physics, astronomy, and other disciplines have necessitated the creation of distributed systems for hosting and processing data. A notable example in this area is the work that Ian Foster and his collaborators have done on developing a generic data grid for high performance computing across large data collections (*10–12*). In industry a key focus area with respect to data management and preparation has been the development of data warehouses (*13*). In one sense, the primary difference between grid computing and a data warehouse is that the former is designed for use and sharing across multiple unaffiliated organizations, while the latter is typically set up to be proprietary to a single organization.

As the quantity of data continues to increase, the data scientist must consider both the data and computational needs. With smaller datasets, data scientists could download the data, and run the analysis locally. With larger datasets we see an increasing need to package the analytical techniques such that the computation can be take place where the data are stored. Such a strategy reduces data traffic, but requires a drastic rethinking of where information resources should be placed. Cloud computing and data storage centers are starting to fill this gap, but costing models continue to evolve along with the technology.

Although the phases of the data lifecycle described above make an alternative perspective apparent, many current discussions of data science focus exclusively on the analytics phase (*14*). As noted in the introduction, although statistical analysis is a mainstay of structured data analytics, to an increasing extent the transformation, assessment, and summarization of non-numeric data is becoming increasingly important (*15, 16*). Within the sphere of non-numeric data, we may encounter both structured data, such as XML documents, and unstructured data, such as natural language text. As one might expect, the difficulty of extracting meaningful patterns from unstructured data is generally more difficult than from structured data. Yet because the amount of unstructured data greatly exceeds the amount of structured data (and continues to grow rapidly), the utility of appropriate unstructured data analysis can sometimes be concomitantly higher (*17*).

Visualization and presentation are intimately connected aspects of data analysis that deserve separate treatment, again because of the ubiquity of data that are not susceptible to statistical analysis. While statisticians have spent decades developing effective numeric and graphic summaries of statistical data sets, the proliferation of non-numeric and unstructured data has created new challenges in visualization and presentation (*18–20*). In addition, effective presentation of data is as much (or more) about effective communication as it is about visualization tools and algorithms. The capability of wrapping a sensible and accurate story around summaries of data arguably requires as much art as it does science; teaching students to become effective presenters can provide more of a challenge than teaching more technical skills.

The phases in the data lifecycle that we have described so far may not have a linear progression (*21*). For example, analyses and visualization of data may uncover anomalies that require returning to the preprocessing phase to conduct more data cleaning, or may even necessitate the collection of additional data. When these cycles settle down, however, the data eventually reach a quiescent stage where no other immediate analysis needs to occur. At this point in the data lifecycle, archiving and preservation become the key activities. Of course, under ideal circumstances the period of quiescence may be quite brief, as an effective data archive makes data available for repurposing and reuse. As an example, consider the Protein Data Bank, an archive of three-dimensional data of biological macromolecules (*22*). The Protein Data Bank serves an international community of biologists and other scientists with a complex archive of more than 80,000 molecules. The archive contains extensive metadata about each molecule in addition to the essential data describing its structure. The archive provides an extensive suite of tools (mainly open source) for adding new molecules into the archive and is organized to facilitate exploration and reuse by researchers.

From a big picture perspective, the phases of the data lifecycle correspond to sets of skills that overlap to some degree with one another, but that also have distinctive identities. An individual who can establish an effective data warehouse or scientific data repository may not necessarily be good at statistical analysis or text mining. It seems possible that one could become a generalist capable of managing projects that span data collection, preprocessing, analysis, and archiving, but the more likely scenario is that one could specialize in one part of the lifecycle while also having workable knowledge or skills in one or more additional aspects. In either case, an interdisciplinary approach to educating data scientists can help bridge the different specialties (*23*). In the next section, we describe several of the base disciplines that would need to be integrated as the basis of such an interdisciplinary approach.

The Ingredients

One factor that differentiates interdisciplinary data science from monodisciplinary approaches to data is the focus on the intersection between the human and technological dimensions of data. The "pagerank" algorithm developed by Google illustrates this difference nicely. Many early search algorithms focused on a statistical analysis of the content of each individual indexed document. In contrast, the innovation that Google brought to the problem was a consideration of how richly a document was linked (particularly inbound links) with the rest of the web. The consideration of the implicit behavioral choices of other web designers was critical in providing an improved indication of the importance of a document. Understanding the behavior and preferences of information users and decision makers who use data is a critical aspect of success as a data scientist (*24*).

This concern for the needs data user dispels one of the criticisms frequently leveled at data science that it, "is not really a new phenomenon" (*25*, *26*). Although each of the core skills that a data scientist must possess has existed in other professions and disciplines for some time, the need for having analytics, infrastructure, and curation knowledge together with a sensitive understanding of the needs of users in a particular application domain has not previously arisen. In the discussion section we will give further consideration to this as a novel feature of data science. Before undertaking that analysis, however, we should make an inventory of the disciplines that have already existed for half a century or longer and whose essential skills and areas of inquiry clearly serve as some of the basis for data science.

First, consider that applied statistics as a discipline is more than a century old, and that statisticians and mathematicians have been devising ever more sophisticated methods of applied analysis since the days of Karl Pearson and R. A. Fisher (*27*). Generally, we regard the applied statistician as a reliable source of expertise on many, though not necessarily all, areas of quantitative analysis. (One exception, for example, is in the application of non-stochastic numerical methods in engineering and science.) Individuals with statistical training can also provide support for the development and interpretation of simulations. Statisticians often have some light programming duties, particularly around the manipulation of data to prepare it for analysis.

For more difficult programming tasks, we generally look toward computer science. A much younger discipline than statistics, computer science as a formal educational area is generally tracked back to 1962, with Purdue University's establishment of the first Department of Computer Science. Although software engineering is just one facet of computer science, for most laypeople the computer science degree is the most noteworthy entry point into a job as programmer or software developer. Interestingly, data science often relies on components of data mining and machine learning; these are areas where computer scientists have also done a considerable amount of research and development. Data mining represents a touch point between computer science and statistics, but computer scientists' use of computational methods to advance this area contrasts with the traditional theoretical methods used in statistics (*28*).

Of course data mining would not be practical without contemporary computing infrastructure and for this we look to the area of information technology, and, for more basic developments, computer engineering. Researchers and practitioners in these areas provide the essential network, storage, and computational power that makes advanced analytics and data visualization possible. The continuing success of the well-known Moore's law is attributable to the rapid advances in capacity and cost effectiveness of the essential cyberinfrastructure that computer engineers and information technologists develop. Relatedly, the worldwide expansion of standardization efforts (e.g., the TCP/IP model maintained by the Internet Engineering Task Force) has helped to ensure the interoperability of large scale cyberinfrastructure (*29*).

In some ways, the subdiscipline of business intelligence incorporates aspects of all three of the areas described above – statistics, computer science, and information technology (*30, 31*). Business intelligence incorporates a mixture of applied statistics, computer science, and information systems to focus on data mining, data analytics, and data visualization applications in the business environment. Although some researchers consider business intelligence as a subset of the forty-year-old area of decision support systems, a Google Ngram viewer search of the phrase "business intelligence" shows a strong uptick starting in 1995 (*32*), while a review of the research literature in decision support systems shows a decline in the production of papers beginning in 1994. For the purposes of this paper we will consider business intelligence as the predominant term (*33*).

Perhaps the most frequently overlooked and underrated contributor in the conversation about data science is the field of library and information science (*23*). Librarianship as a field predates statistics, and contemporary educational programs of library and information science (LIS) predate both information technology and computer science (*34*). The curatorial component of data science is consistent with the traditional mission of librarianship to maximize the "effective use of graphic records" (*35*), and the aims of the information professions more broadly to add value in alignment with the needs of user groups (*36*). More specifically, a core LIS role focuses on providing information services for scientists and scholars through the representation, preservation, organization, and management of scientific and scholarly products (*37*). As is the case with data science more generally, the evolving data curation knowledge base in LIS draws on cognate disciplines, especially archival science, computer science, and domain based informatics.

A second reason may be that some of the highly technical components of LIS – such as text mining – have not heretofore received the same level of attention as mainstream analytical and data mining techniques such as regression (*38*).

Yet, without methods and standards developed by or in close collaboration with LIS scholars and practitioners (e.g., the Dublin Core), many of the scientific and engineering data sharing practices currently in existence would not be possible (*39*).

Taken together, these five disciplines – statistics, computer science, information technology, business intelligence, and library and information science, comprise the core sources of knowledge and skill that appear frequently in discussions of data science (*40*). Likewise, in analysis of the tasks and activities of professionals in eScience, a mixture of skills from these five disciplines covers much or all of the necessary background (*41*). Finally, in a review of available jobs, the qualifications of individuals educated in one or more of these majors seems to satisfy the expected requirements for many available positions documented in the jobs marketplace (*42*).

Results of the Workshop

Overview

To make progress on defining the educational challenges of data science, the authors led a brainstorming workshop focused on specifying appropriate terminology and focus areas of data science. The workshop was part of the program for the seventh annual iConference that occurred in Toronto, Canada in February 2012. The iConference – sponsored by a consortium of more than 35 international schools of information – serves as the primary venue for interdisciplinary scholarship in the allied information fields. The authors of the present chapter submitted for peer review and received an acceptance for a workshop that planned to bring together information experts from academia and industry to discuss the educational dimensions of data science. The workshop had attendance of slightly more than 20 experts. The experts included faculty researchers, deans of information schools, information industry advocates, and doctoral students in information programs. Following a brief plenary session, the participants broke into four brainstorming groups, each addressing a different dimension of data science education. Simultaneously, group facilitators created notes and sketches on poster-sized easel pads for later use. After the completion of the breakout groups, another unified discussion period provided an opportunity for facilitators to synthesize the group outputs. Doctoral student note takers recorded the results of the breakout sessions and the synthesis session into the common Google document. Analysis of the resulting notes provided the summary that appears below.

Defining the Four Groups

While planning the workshop, the authors considered the results of a 2011 Research Data Workforce Summit (*23*) as they formulated the goals for the session: to sharpen the identity of professionals in data science, to consider key educational challenges, and to develop a shared vocabulary of key responsibilities in the education of data science professionals. The authors identified four distinct topical areas that merited separate treatment during the breakout component of the session. The four groups were data curation, data analysis, cyberinfrastructure, and domain knowledge.

The data curation group was charged with considering the data lifecycle from the point of data collection through reuse and into archiving and preservation. We entitled this group the "curation" group. The second group was concerned with the reuse phase for the data, and in particular the pre-processing, transformation, analysis, and visualization of data for use by decision makers, researchers, and other data end users. We entitled this group the "analytics" group. The third group focused on infrastructure and particularly on the necessary computing, software, and network technology that would facilitate data sharing and reuse across functional user areas. This group had the title "networks and infrastructure." Finally, the fourth group had the task of considering questions surrounding domain knowledge. The domain knowledge group considered a long standing issue for the information professions and other "meta-fields" (i.e., professions that cross cut other disciplines), namely that information professionals must immerse themselves in domain knowledge in order to have a sufficient understanding of the problems and questions their work is trying to address.

During the kickoff plenary session each group received instructions concerning the three goals of the workshop: sharpening identity, enumerating educational challenges, and establishing common vocabulary. Some natural variation occurred in the priorities given to these three areas, such that the groups sometimes provided dissimilar, yet complementary results. In the material below, we first present a narrative overview of each of the group discussions, focusing attention on one or two key points for each group. Following the narrative overview, we synthesize the results into a single summary that integrates the four areas.

Group Discussion Overviews

The curation group tackled the question of overlap between different specialty areas in the data lifecycle. Working backwards through the lifecycle, the group noted that archiving concerns the long-term storage of data for future research, while curation is viewed as a process that adds value throughout the data lifecycle. At a more micro level, data management, while bearing some similarities to curation, was viewed as a localized activity performed for particular users or user groups. A key point of consensus was that data curation was primarily concerned with the idea of "fit for purpose." By applying standards and best practices, accurately representing the provenance of data, and by preparing data for reuse, data professionals can provide information users with assurances that the data they rely upon is fit for re-use or new uses. This consensus led to discussion of a subsidiary concept of a "data broker" – a professional who mediates between the sources of data and the users of data to provide these assurances, but a consensus did not form around the adoption of this term.

The analytics group focused on developing an informal taxonomy of analytics task areas and on creating a list of specific skills required to perform those particular task areas. The key areas fell into four key areas: (1) the ability to connect between a problem domain and how to frame the problem such that computational techniques could be used (2) data representation including information extraction, feature extraction, graphical representations (3) data analytics such as text mining, statistics, data mining, social network analysis and (4) data presentation. The group then prioritized the list of activities to differentiate between core skills and secondary skills necessary for a data scientist. Interestingly skills such as programming, databases, and pattern recognition were deemphasized.

The networks and infrastructure group confronted an essential challenge of developing and applying the technologies necessary to support data science. Data science is dependent on extensive infrastructure ranging from grid computing to networks to distributed storage, but data science expertise may not realistically include deep knowledge of these many technologies. Similarly, while a data scientist may have extensive knowledge of certain application software (such as statistical analysis software), it is unlikely that this knowledge will extend into core operating system and middleware areas. The group's response to this paradox was to focus data science on the communicative and managerial aspects of infrastructure. Specifically, the group identified requirements analysis, user-centered design, and project management as key areas where data scientists could be expected to contribute to the development and maintenance of technology infrastructure.

In the domain group, members wrestled with a somewhat different conflict, the balance of specialty skills unique to data science versus domain specific knowledge needed to thrive in a particular application area. As is the case with many areas of study, there is a practical upper limit on how much domain knowledge a student can absorb. For example, a data scientist who facilitates chemical informatics for a genomics firm or a group of academic researchers might have undergraduate coursework in chemistry, a specialized graduate course in chemical informatics, an internship in this area, or all three. Even with all three, however, domain-specific challenges are likely to arise in areas that have not been covered by the data scientist's formal education. Although limitations in the length of the program of study make this a difficult problem to resolve, the group believed that it was the responsibility of educators to unearth commonalities in data management practices across application domains and to highlight these in a way that provided students with appropriate schema for future learning. For example, in the area of data governance different norms may predominate in the academic research sector than in the industrial research sector because of their different orientations to data sharing. Identifying and analyzing such differences in data governance would make it possible for students to learn domain specific norms more quickly in the future, regardless of the sector in which they eventually worked.

Synthesis of Group Results

At the conclusion of the breakout sessions summarized above, the workshop participants came back together to synthesize the topics and perspectives generated by the individual groups into a coherent whole. In this section we present this basic holistic picture, providing a more thorough discussion of the implications in the final section of the chapter.

The synthesis discussion converged on a the idea introduced by a workshop participant, the "T-shaped professional." Several sources attribute the first use of this phrase to a 1991 article in *The Independent* (a London newspaper) by freelance author David Guest, but a thorough search of newspaper archives did not turn up that article. Nonetheless, IBM and other firms have extolled the value of an education that produces T-shaped professionals – individuals with a wide base of knowledge across subject areas along with a deep expertise in one area – represented by the wide top of the T and its long descender (*43*).

In application to data science, the group discussed several possible modifications to the T-shaped notion. First, participants agreed that the top of the T should comprise the areas represented by the first three breakout groups: data curation; analytics, visualization, and presentation; and networks/infrastructure. Deep knowledge in one of these three areas could constitute the descender of the T. Table 1 lists sample task activities enumerated for each of the three areas.

Table 1. Sample Tasks in Three Data Science Expertise Areas

Data Curation	*Analytics/Vis./Pres.*	*Networks/Infrastructure*
Provide data access	Preprocess data	Assess requirements
Facilitate data deposits	Transform data	Develop services
Manage representations	Integrate data	Integrate software
Ensure interoperability	Analyze data	Evaluate solutions
Archive/preserve data	Mine data	Manage projects
Plan curation strategies	Evaluate/interpret results	Talk with developers
Establish data policies	Visualize results	Design databases

The tasks appearing in Table 1 represent just a few examples of activities appropriate to each of the three specialty areas. More detailed job analysis work would be necessary to ensure full coverage of the range of tasks expected of in a data scientist role in three areas, but the table demonstrates that a coherent set of three specialty areas can be delineated for the top of the T and that data scientists might choose to specialize in one of the three areas.

The T-shaped concept accounts for both broad knowledge of a practice area such as data science as well as the possibility of specialization. In addition, the group agreed that data science skills and knowledge require some essential understanding of the application domain. As noted above, a data scientist who works in the area of chemical informatics needs knowledge of chemistry in addition to a base of knowledge in informatics. In fact, one of the workshop outcomes was the recognition that for data science the T-shaped model could be extended to an "I-shaped" model by adding a broad (but probably not deep) base of domain knowledge across the bottom. In this respect, I-shaped expertise would encompass a broad understanding of techniques across data science; a deep specialization in curation, analytics, or infrastructure; and a general level of knowledge of at least one application domain.

Discussion

The literature review presented above, together with the results of the workshop, paint a picture of some of the necessary requirements for educating data scientists. A data scientist must have a basic understanding of data curation, data analytics, infrastructure, and the data domain. To be successful, a data scientist needs to develop deep expertise in at least one of these areas, but the depth of expertise across areas may vary greatly. The data scientist role emphasizes the value for information users and decision makers that can come about through application and innovative use of existing technology to organize, analyze, and curate data. Focusing on the contextual value of computing and the data analysis it supports is an orientation that has been labeled "computing with a purpose" (*44*). This orientation meaningfully distinguishes data science from monodisciplinary approaches to data. A statistician is trained to create and apply statistical analysis methods. A computer scientist is trained to create and apply algorithms and computing infrastructure. In contrast, a data scientist is trained to *serve the needs of information users* by melding analytics, infrastructure, and curation. One unintended but welcome side effect of this orientation may be the possibility of increased diversity: For example, Tracy Camp's study addressing the low numbers of women in computer science suggested that educational computing projects with prosocial goals helped reverse the decline (*45, 46*).

This point highlights one key set of skills that workshop participants discussed, but that were not reflected in the composition of the four break-out groups: A data scientist needs interpersonal and communication skills to be successful. Knowing how to elicit information about the data collection processes when they are not clearly documented, understanding how to present information such that others can reuse data, and being able to learn new technologies as they emerge: these are all core skills of the successful data scientist.

With respect to technology skills, the workshop participants suggested that while essential understanding of cyberinfrastructure was important for data scientists, the necessary levels of skills and knowledge were notably different than those required for computer scientists and computer engineers. There was certainly a time when the transformation and analytics activities required by a data scientists required an in-depth background in computer science. However, advances in storage and analytical tools now make many of these activities accessible to users who have more modest levels of technical training.

A Modified Continuing Education Model

The discussion of an "I-shaped" education in the workshop session underscored some of the difficulty of gaining deep knowledge in every area of data science. This depth versus breadth problem comes up in many professions such as medicine and psychology, where the number and extent of specialty areas is far too large for any one person to master. In any technology intensive field, such as data science, the rapid advances in software and hardware that drive changes to standards and tools compound this problem.

To address these concerns, we recommend a strategy based on the continuing education model used in many professions. This continuing education model fits perfectly with the "I-shaped" education that workshop participants discussed. Specifically, we suggest that the education of a data scientist should begin with learning in one of the areas represented by the descenders of the I. Any one of these educational options could work:

- An undergraduate degree in information technology or an allied information field such as information management, network management, or telecommunications (*47*)
- An undergraduate degree in computer science or computer engineering
- An undergraduate degree in applied statistics or mathematics with a concentration in statistics
- A graduate degree in library and information science (undergraduate degrees in this field are rare)

Next, the novice data scientist can obtain domain knowledge through an internship, through on the job experience, or alternatively with additional formal education (e.g., a minor in chemistry for an individual in chemical informatics). Among these choices, the internship experience will likely provide the greatest job opportunities upon graduation from a degree program, whereas on the job experience can provide the quickest and most intensive learning curve for obtaining domain knowledge. The key recommendation here is that obtaining domain skills and knowledge is best accomplished through a work-based learning strategy rather than primarily through formal education.

Finally, the data scientist, now armed with a specialization in curation, analytics, or cyberinfrastructure, can strengthen their knowledge in the other two areas of data science through continuing education (CE) experiences. Opportunities to obtain these CE experiences could arise from professional conferences, through certificate programs offered online or face-to-face by universities, or through part- or full-time graduate work in a professional master's degree program that provides a data science curriculum. Generally speaking, these certificate degree programs have a focus on one of the three areas of data science (curation, analytics, or infrastructure), and the data scientist should seek complementary education in the area where he or she has the least prior knowledge (and the most professional need). Examples of such programs available at the time of this writing include:

- North Carolina State University – Master's of Science in Analytics
- Northwestern University – Master's of Science in Predictive Analytics
- Stanford University – Graduate Certificate in Data Mining and Applications

- Stevens University – Master's of Science in Business Intelligence and Analytics
- Syracuse University – Certificate of Advanced Study in Data Science
- University of California at San Diego – Graduate Certificate in Data Mining
- University of Illinois – Specialization in Data Curation (development of a specialization in Sociotechnical Data Analytics is underway)

Note that each of these programs has a distinctive focus. For instance, a library and information science graduate who was already strong in curation, might choose to broaden their knowledge and skills in analytics by undertaking UCSD's graduate certificate in data mining. In contrast, an individual with an undergraduate degree in statistics might increase their breadth by studying in the Data Curation specialization at the University of Illinois. Undoubtedly, additional programs with other mixes of specialization will emerge over time and provide new options for distance education and/or part time study for working professionals.

Next Steps

The present chapter provided background information and an overview of the contributions of a group of roughly 20 experts from the information field. The generality of these contributions is naturally limited by the backgrounds of individuals who participated in the workshop. Although the workshop participants came from many different disciplines, the mixture may not have fully represented the perspectives of computer scientists, statisticians, library scientists, business scholars, and others who may have professional interest in data science.

A more comprehensive and generalizable effort should use the techniques of job analysis to obtain and integrate input from a representative sample of subject matter experts from science and industry (*41*). The results might shed light on a range of details that the present chapter was not able to uncover, e.g., the appropriate educational balance between the different areas of data science; additional subject areas that were inadequately represented in this analysis; and/or alternative educational approaches that could supplement or replace the proposed continuing education model.

Keeping in mind these limitations, the primary contribution of the present chapter was to define a high level structure for thinking about data science education – the breakdown between curation, analytics, cyberinfrastructure plus domain knowledge – and the recommended modified continuing education model. Combining the "I-shaped" notion with a mixture of formal and work-based learning can provide an optimal path to educating data scientists for productive roles in various work settings. In addition to this primary contribution, we also identified a critical differentiating factor that sets the multidisciplinary data scientist apart from monodisciplinary specialists: the data scientist uses data curation, data analytics, and cyberinfrastructure to serve the data needs of information users and decision makers.

References

1. Allard, S.; Levine, K. J.; Tenopir, C. Design engineers and technical professionals at work: Observing information usage in the workplace. *J. Am. Soc. Inf. Sci. Technol.* **2009**, *60* (3), 443–454.
2. Blake, C.; Pratt, W. Automatically Identifying Candidate Treatments from Existing Medical Literature. In *AAAI Spring Symposium on Mining Answers from Texts and Knowledge Bases*, 2002.
3. Palmer, C. L. *Work at the Boundaries of Science: Information and the Interdisciplinary Research Process*; Kluwer: Dordrecht, 2001.
4. Stanton, J. M.; Rogelberg, S. G. Using internet/intranet web pages to collect organizational research data. *Organ. Res. Methods* **2001**, *4* (3), 200–217.
5. Hey, A. J. , S. Tansley; Tolle, K. M. *The Fourth Paradigm: Data-Intensive Scientific Discovery*; Microsoft Research: Redmond, WA, 2009.
6. Manyika, J.; et al. Big Data: The Next Frontier for Innovation, Competition, and Productivity; McKinsey Global Institute, 2011.
7. Hey, T.; Trefethen, A. E. The UK e-Science Core Programme and the Grid. *Future Gener. Comput. Syst.* **2002**, *18* (8), 1017–1031.
8. Atkins, D. Revolutionizing Science and Engineering through Cyberinfrastructure: Report of the National Science Foundation Blue-Ribbon Advisory Panel on Cyberinfrastructure; National Science Foundation: Arlington, VA, July 9, 2003.
9. Kohavi, R.; Mason, L.; Parekh, R.; Zheng, Z. Lessons and challenges from mining retail e-commerce data. *Mach. Learn.* **2004**, *57* (1), 83–113.
10. Chervenak, A.; Foster, I.; Kesselman, C.; Salisbury, C.; Tuecke, S. The data grid: Towards an architecture for the distributed management and analysis of large scientific datasets. *J. Network Comput. Appl.* **2000**, *23* (3), 187–200.
11. Ranganathan, K.; Foster, I. Identifying Dynamic Replication Strategies for a High-Performance Data Grid. *Grid Compututing—GRID 2001*, 2001, pp 75–86.
12. Foster, I.; Kesselman, C. *The Grid: Blueprint for a New Computing Infrastructure*; Morgan Kaufmann Publishers: Burlington, MA, 2004.
13. Inmon, W. H. *Building the Data Warehouse*; Wiley: New York, 2005.
14. Cohen, J.; Dolan, B.; Dunlap, M.; Hellerstein, J. M.; Welton, C. MAD skills: New analysis practices for big data. *Proc. VLDB Endowment* **2009**, *2* (2), 1481–1492.
15. Bose, R. Advanced analytics: Opportunities and challenges. *Ind. Manage. Data Syst.* **2009**, *109* (2), 155–172.
16. Keim, D. A.; Kohlhammer, J.; Ellis, G.; Mansmann, F. *Mastering the Information Age: Solving Problems with Visual Analytics*; Eurographics Association: Goslar, Germany, 2010.
17. Greengrass, E. Information Retrieval: A Survey, 2000.
18. Ahern, S.; Naaman, M.; Nair, R.; Yang, J. H. I. World Explorer: Visualizing Aggregate Data from Unstructured Text in Geo-Referenced Collections. In *Proceedings of the 7th ACM/IEEE-CS Joint Conference on Digital libraries*, 2007; pp 1–10.

19. Feldman, R.; Sanger, J. *The Text Mining Handbook: Advanced Approaches in Analyzing Unstructured Data*; Cambridge University Press: Cambridge, UK, 2007.
20. Silver, D.; Wang, X. Tracking Scalar Features in Unstructured Data Sets. In *Visualization '98. Proceedings*, 1998; pp 79−86.
21. Fayyad, U.; Piatetsky-Shapiro, G.; Smyth, P. The KDD process for extracting useful knowledge from volumes of data. *Commun. ACM* **1996**, *39* (11), 27–34.
22. Berman, H. M.; et al. The protein data bank. *Nucleic Acids Res.* **2000**, *28* (1), 235–242.
23. Varvel, V. E.; Palmer, C. L.; Chao, T.; Sacchi, S. *Report from the Research Data Workforce Summit*; Champaign, IL, 2011.
24. Blake, C.; Pratt, W. Collaborative information synthesis I. A model of information behaviors of scientists in medicine and public health. *J. Am. Soc. Inf. Sci. Technol.* **2006**, *57* (13), 1740–1749.
25. Beware the Hype over Big Data Analytics, Seeking Alpha, 2012. [Online]. http://seekingalpha.com/article/441171-beware-the-hype-over-big-data-analytics.
26. Mitchell, R. L. Hadoop Hype and Data Yodas: Tales from Predictive Analytics World. *Computerworld*, 2012
27. Stanton, J. M. Galton, Pearson, and the peas: A brief history of linear regression for statistics instructors. *J. Statistics Educ.* **2001**, *9* (3), 1–13.
28. Banzhaf, W.; Nordin, P.; Keller, R.; Francone, F. *Genetic Programming: An Introduction on the Automatic Evolution of Computer Programs and Its Applications*; The Morgan Kaufmann Series in Artificial Intelligence; Morgan Kaufmann Publishers: Burlington, MA, 1997.
29. Pipek, V.; Wulf, W. Infrastructuring: Toward an integrated perspective on the design and use of information technology. *J. Assoc. Inf. Syst.* **2009**, *10* (5), 447–473.
30. Golfarelli, M.; Rizzi, S.; Cella, I. Beyond Data Warehousing. In *Proceedings of the 7th ACM International Workshop on Data Warehousing and OLAP - DOLAP '04*, 2004; p 1.
31. Negash, S. Business Intelligence. *Commun. ACM* **2004**, *13*, 177–195.
32. Power, D. J. Decision Support Systems: A Historical Overview. In *Handbook on Decision Support Systems*, Volume 1; Springer: Berlin, Heidelberg, 2008, pp 121−140.
33. Arnott, D.; Pervan, G. A critical analysis of decision support systems research. *J. Inf. Technol.* **2005**, *20* (2), 67–87.
34. Wiegand, W. A. Tunnel Vision and Blind Spots: What the Past Tells Us about the Present. Reflections on the Twentieth-Century History of American Librarianship. *The Library Quarterly*, 1999, pp 1−32.
35. Shera, J. H. *The Complete Librarian and Other Essays*; Press of Case Western Reserve University: Cleveland, OH, 1971; p 57.
36. Taylor, R. S. *Value-Added Processes in Information Systems*; Ablex: Norwood, NJ, 1986.

37. Palmer, C.; Renear, A. H.; Cragin, M. H. Purposeful Curation: Research and Education for a Future with Working Data. In *Proceedings of the 4th International Digital Curation Conference*, 2008.
38. Blake, C.; Pratt, W. Better Rules, Fewer Features: A Semantic Approach to Selecting Features from Text. In *Data Mining, 2001. ICDM 2001, Proceedings IEEE International Conference on*, 2001; pp 59−66.
39. Miller, E. An introduction to the resource description framework. *Bull. Am. Soc. Inf. Sci. Technol.* **2005**, *25* (1), 15–19.
40. Kim, Y.; Addom, B. K.; Stanton, J. M. Education for escience professionals: Integrating data curation and cyberinfrastructure. *Int. J. Digital Curation* **2011**, *6* (1), 125–138.
41. Stanton, J. M.; et al. Education for escience professionals: Job analysis, curriculum guidance, and program considerations. *J. Educ. Libr. Inf. Sci.* **2011**, *52* (2), 79–94.
42. Addom, B. K.; Kim, Y.; Stanton, J. M. eScience Professional Positions in the Job Market. In *Proceedings of the 2011 iConference. iConference '11*, 2011; pp 630−631.
43. Bullen, C. V.; Abraham, T.; Gallagher, K.; Simon, J. C.; Zwieg, P. IT workforce trends: Implications for curriculum and hiring. *Commun. Assoc. Inf. Syst.* **2009**, *24* (9), 129–140.
44. Margolis, J.; Fisher, A. *Unlocking the Clubhouse: Women in Computing*; MIT Press: Boston, 2002; pp 49−60.
45. Camp, T. Women in computer sciences: Reversing the trend. *Syllabus* **2001** (August), 24–26.
46. Camp, T. The incredible shrinking pipeline. *Commun. ACM* **1997**, *40* (10), 103–110.
47. Stanton, J. M.; Guzman, I. R.; Stam, K. R.Information Nation: Education and Careers in the Emerging Information Professions. *Information Today, Inc.*, 2010, p 256.

Chapter 7

Data Management Services in Libraries

Patricia Hswe*

Publishing and Curation Services, University Libraries,
The Pennsylvania State University, W-311 Pattee Library,
University Park, Pennsylvania 16802
*E-mail: phswe@psu.edu

Activated in January 2011, the National Science Foundation requirement to include data management plans with grant proposal submissions has compelled many academic libraries to revisit their service offerings, particularly in the context of assisting researchers with managing their data in efficient, productive ways. This chapter covers the status of data management services in academic libraries in the years since the NSF requirement went into effect. It explores the reasons why data need to be managed, and why librarians and libraries are up to the task; describes the steps that various libraries have taken to determine service, staffing, and infrastructure requirements; provides a brief overview of current data management service offerings in libraries; and, in conclusion, touches on the emergence of new librarian roles, including postdoctoral positions, that have arisen to meet demands in data management and data curation in academic libraries.

Introduction

It's the 21st century - do you know where your data are?

Research data exemplify the output of experimentation. As such, data can be considered "the recorded factual material commonly accepted in the scientific community as necessary to validate research findings" (*1*). When the National Science Foundation (NSF) announced in summer 2010 that, effective

© 2012 American Chemical Society

January 18, 2011, data management plans (DMPs) would be required as part of grant application submissions to the agency, it may as well have been asking the question above. Data also tell a story, one that the NSF - which applies taxpayer dollars toward an estimated 20% of all research funded by the U.S. government - wants to ensure that scientists are recounting as fully as possible. The general guidance on DMPs provided by the NSF is intended to orient researchers with the essentials of managing data, even prior to their generation, when methodologies and other approaches are being determined, through their collection and continuing into their life cycle of use and relevance, with the overall aim of sharing data and making them publicly accessible. To this end, in particular for the directorates that have not issued more specific DMP guidelines, such as the Computer & Information Science & Engineering Directorate and the Division of Material Research (in the Directorate of Mathematical and Physical Sciences), the NSF outlines the components that make up a plan, which amount to five sections:

- a description of the types of data;
- how the data will be documented (i.e., metadata standards to be used);
- policies for sharing the data and providing access to them (also accounting for any confidentiality, privacy, or security protections);
- guidance and recommendations for reuse and redistribution of the data, as well as creation of derivatives from them;
- how the data will be archived, to enable preservation for continuing access.

These are the plot points necessary for telling the story of research data.

Although library literature has long acknowledged the "data deluge," and calls for librarians to acquire skills in data curation and data management were appearing before the NSF announcement, this recent mandate has many academic libraries, whether large or small, marshaling resources toward new service models, new infrastructure, and new librarian roles (2–10). In addition, in the short time following this new implementation of the NSF's data sharing policy, more library and information science degree programs are integrating courses that address research data management issues, including short-term, "boot camp" courses, or they are including coverage on this topic in already existing classes (11). Fellowships in data curation, with the goal of expanding a workforce of expertise in this area, have also emerged since the NSF went public with its requirement (12).

This chapter covers the status of data management services in academic libraries in the years since the NSF requirement went into effect. It explores the reasons why data need to be managed, and why librarians and libraries are up to the task; describes the steps that various libraries have taken to determine service, staffing, and infrastructure requirements; provides a brief overview of current data management service offerings in libraries; and, in conclusion, touches on the emergence of new librarian roles, including postdoctoral positions, that have arisen to meet demands in data management and data curation in academic libraries.

Why Data Need To Be Managed and the Key Role Librarians and Libraries Play

The story of data management services in libraries cannot be told without considering the subtext of the story first: the unstoppable torrent itself of data. The escalating availability to researchers of rapid and scalable computational methods, afforded by highly efficient and powerful supercomputing processors, has produced an ineluctable embarrassment of data riches. As Hay et al explain, scientists are struggling with a massive amount of data born of many types of sources, including instruments, simulations, and sensor networks (*13*). As a consequence, *not* managing data properly has come at great price to many scientists and scholars. Reports of data loss, security breaches, and lack of data authentication are not uncommon and ultimately erode both the sharing of data and access to them. The steepest cost is, likely, impeded progress in scientific research: the more time scientists must spend on management of data, the less time they have to do the research that new data and future findings are dependent upon.

Perhaps most significant, the mismanagement of data diminishes the reproducibility and replicability of data - those verification processes that constitute the gold standard in scientific research. A special issue of *Science* addressed the topics of data replication and reproducibility, noting an array of challenges and benefits, such as the following: processes observed in the field that call for certain parameters or conditions are difficult to capture and thus replicate; the discipline of computer science needs a set of base norms, or standards, for reproducibility of methods and code, so that researchers can arrive at identical, dependable outcomes with the unprocessed, initial data; and allowing for different groups, such as scientists and public policy makers, to view the same data - and engender new data - can yield rich results (*14*). This last point resonates with what José Muñoz, Chief Technology Officer (CTO) in the NSF's Office of Cyberinfrastructure, stated in the May 10, 2010, press release announcing the funding agency's imminent requirement: "Twenty-first century scientific inquiry will depend in large part on data exploration" (*15*). Because of the mandate, such data exploration can be enabled across communities more frequently, spurring new findings and - by extension - new stories and new questions.

One of these communities is, arguably, that of academic librarians and libraries. Libraries have long been in the practice of maintaining, as well as disseminating and sharing, the scholarly record, of which research data constitute an intrinsic part. Starting up new library services for data curation presents an opportunity for librarians who are subject experts to collaborate with archivists, who are experienced in the appraisal, selection, description, organization, preservation, and retention of a range of content types, from electronic mail, to scholarly papers, to university business records (*16*). As Heidorn asserts, "Curation of data is witihin the libraries' mission, and libraries are among the only institutions with the capacity to curate many data types" (*17*). At the same time - while librarians possess the subject expertise - for some, as Heidorn also points out, becoming conversant with data management practices presents a formidable learning curve (*17*). It necessitates a working knowledge of not only basic archival approaches but also metadata standards, intellectual property issues, institutional research administration policies and guidelines, and conventions in file management and storage practices.

Working with researchers on management of their data may seemingly cast librarians in roles that are different from before, but, in effect, this interaction coheres well with what librarians do and have always done. For example, liaison librarians - who typically have subject expertise - develop and manage collections driven by the research that engage their faculty and students. They are also trained in conducting reference interviews. As Witt and Carlson suggest, liaison librarians could effectively merge these approaches through an activity they call a "data interview," during which librarians identify and characterize data sets faculty have generated as part of their research (*6*). In a departure from the typical targets of collection management, such as monographs and serials, the indicated sets of data in effect become "information assets to be collected, preserved, and made accessible as a function of the library's collection development" (*6*). To launch the interview, Witt and Carlson recommend asking researchers, "What is the story of your data?" The process of librarians working with researchers to figure out the story of their data is an investigative activity – akin to the "data exploration" that CTO Muñoz, of the NSF, advocates is pivotal to present-day scientific inquiry. In addition, collaborating with researchers in this capacity means that acquisition of published work cannot remain the key goal in collection development; ensuring procurement of related data files, as well as of software applications that the data are dependent upon for readability and analysis, emerge as significant objectives (*4*).

In what other ways are librarians prepared to undertake the development and implementation of research data management services? With the diversity of domains and disciplines represented at any given institution, how will librarians, many of whom individually serve multiple departments and thus myriad faculty research interests, address data management needs through scalable, sustainable approaches? What kinds of collaborations will need to be fostered, toward building effective service models for helping researchers curate their data? What role, if any, does institutional commitment play - perhaps in the form of institutional policies that not only support research data management practices but also even require them? How can librarians prepare graduate students, who are the next generation of researchers, to be data literate, toward gaining practical experience in data curation methods? These questions are not easily answered, but paths to solutions have begun to be paved, a few of which are considered next.

Paths to Providing Data Management Services

As Salo advises, bringing libraries up to speed to accommodate researcher needs in data management will demand a "retooling," toward an understanding of the following: what research data are (e.g., their extent and scale, their project-based context, their heterogeneous nature, and the non-standard formats that data can take); and whether the functionalities of digital libraries and institutional repositories - the mainstays for digital content in most libraries -

map, or not, to the life-cycle management of data (*18*). The goal of the retooling process is to align, as much as possible, a library's technical infrastructure and services with what researchers require in order to manage their data efficiently and effectively. To do this, many librarians have undertaken assessment work, such as surveying their faculty, or targeting specific ones for interviews, about their research data. They have also developed pilot studies to inform the development of service models. And still others have launched self-educational, "inreach" activities for the short-term, such as reading groups and topic-based information sessions, to help gain internal traction with data management issues. Additional literature describing tiers of service and concepts of success in the research data management space - such as collaboration that is transparent and all encompassing and a commitment to learning as a community - have surfaced in the last couple of years as well (*19, 20*).

Below is an overview of some of these diverse paths to new service models and infrastructure for addressing researcher needs in data management and data curation. (Not all of the experiences encapsulated here occurred directly after the NSF mandate, but the earlier timing does not diminish the value or relevance of the understanding that was achieved.)

Librarians are no strangers to survey instruments and semi-structured interviews, the chief methods they have exploited for evaluation of what researchers are doing with their data. In some efforts, a team of librarians develops its own survey instrument, or it draws on existing tools - as Georgia Tech Library did with the Data Asset Framework (*21*), originated by the U.K.'s Digital Curation Centre (*22, 23*). In 2006-2007 the University of Minnesota Libraries created its own survey for a "Science Assessment" study, in which sixteen librarians with subject expertise in science met with more than 70 researchers of varying statuses - faculty members, postdoctoral fellows, and graduate students - to discuss their research practices and needs in focus-group settings and interviews (*7*). A significant question that the librarians asked researchers, all of them scientists, in this study was, "If you seek assistance from the library, what kinds of help are you looking for? What kind of assistance is needed? (For grants? Publishing? Data curation and preservation?)" (*7*). The main areas where respondents sought help were organization and manipulation of data; storage, security, and sharing of data (specifically, the apparent lack of standards and guidance to consult); and stewardship of data - including some skepticism about whether this is even possible or necessary, given the inability to know the value of one's research data years from now.

Evaluating the state of research data management at one's institution can also mean targeting even more specific populations, such as the Primary Investigators (PIs) of NSF-funded projects and projects funded by the National Institutes of Health (NIH). Peters and Dryden did just this at the University of Houston as the basis of a pilot study, contributing to the university's overall impetus to strategize for more robust research funding (*24*). Besides interviewing the PIs, a chief objective of the study was to gather information about data management approaches in the context of the projects the PIs directed, totaling fourteen, which were both group- and individual-based. The team made valuable discoveries pertaining to project information, data life-cycle workflows, data characteristics (e.g., types of data being generated across the projects), data management (e.g., methods for data storage and access), data organization, and data use. Outcomes from the study included a proposal to form a library-based "Data Working Group," to consolidate efforts among liaison librarians and to communicate with researchers in consistent fashion. Plans also evolved to sponsor an event bringing together "data service providers" from various parts of the campus, including the libraries, IT, and research centers, to gather and share what is being accomplished across the university and try to discontinue endeavors that are redundant. More of these types of collaborations will be needed in the near future, if institutions are to align service offerings and policies for data management. In addition, the pilot study team wished to expand their investigation to include researchers in non-science domains as well as researchers in science and engineering who have not secured funding-agency support for their projects.

Goldstein and Oelker undertook an approach at Mount Holyoke College, a small liberal arts college in New England, similar to that of Peters and Dryden at the University of Houston. Beginning in summer 2010, the Library and Technology Services (LITS) Department at Mount Holyoke worked to address the NSF requirement in order to be prepared to help faculty fulfill it, engaging the College's Sponsored Programs Office and various subject librarians. LITS liaison librarians polled faculty about the following: how much research data in digital format they have; how much they expect this data to increase by June 2012; and whether they have lost data in recent years that were not backed up (2). The librarians also enlisted the support of the Associate Dean of Faculty for Sciences to ensure as much cooperation from faculty members as possible. In addition, the head of Digital Assets and Preservation Services, which is responsible for preservation of digital content, began collaborating with a science liaison librarian. Together they formed an early "response team" to address requests from Mount Holyoke faculty for assistance with DMP development (2). Besides surveying their faculty on basic data management practices, LITS librarians sought to find out what might be happening in data management at neighboring institutions, such as the University of Massachusetts at Amherst, to suss out possible external collaborations (25). Based on their data management services start-up experience, Goldstein and Oelker recommend that librarians at small institutions "adopt a policy of cooperation and collaboration" and work proactively to address those "just-in-time" needs for help with DMPs from faculty, while also strategizing for expansion and improvement of existing tools and services, such as, in the case of Mount Holyoke, its DSpace repository instance (2). Goldstein and Oelker also emphasize keeping abreast of what peers are doing in the area of data management and of what is occurring nationally.

In addition to assessing research data management activities on campus (i.e., external to the library), librarians have also been looking within their bounds, performing gap analyses via surveys distributed among information professionals in research libraries. For example, librarians at the University of Massachusetts Medical School did a study to evaluate required competencies for providing e-science research activities, such as data curation and management services. Surveying librarians in six U.S. states (out of 141 librarians who received the survey, 63 responded), Creamer et al determined that while a small percentage of librarians were actively providing such services, more than half the number of respondents were involved in creating a strategic plan or policy for data management (26). Some of the competencies distinguished by Creamer et al in their study were the following: technical competencies, such as providing data archiving and preservation services, working with metadata standards, and managing an institutional repository; and non-technical competencies, such as outreach and instruction in various aims of scholarly communication (data sharing, open access, intellectual property rights, data literacy), conducting data interviews, working with researchers on DMP development, and finding and locating data that their patrons need for their own scholarship. For Creamer et al, these survey findings were instrumental in the development of both the "e-Science Portal for New England Librarians" and a data curation and management curriculum as professional advancement resources (26).

Another tactic employed toward an improved understanding of data management service requirements has been to participate in, and promote, "inreach" activities of the "train the trainer" mode: that is, a librarian or small team of librarians with base knowledge of data management practices organizes instruction sessions, workshops, or reading groups as vehicles for distributing this knowledge among their colleagues. At the University of Virginia, Sallans and Lake formed a "Scientific Data Consulting Group," one of whose primary aims was to conduct bi-weekly "Data Curation Brown Bag" discussions (27). The intent behind these brown bags was threefold: 1) inform subject specialists of urgent topics and trends in data curation; 2) give a brief talk and provide a one-page, so-called white paper summarizing the issue at hand, followed by informal discussion; and 3) enable subject experts to become familiar enough with data curation for them to engage in discussions about it with the departments and faculty whose interests they represent. At the aforementioned University of Houston, Peters and Dryden had plans to develop "data 101 instruction sessions" to help their colleagues conquer the data literacy learning curve (24).

Other institutions have ventured beyond assessment and internal instruction and launched pilot projects to test proof-of-concept aims. In spring 2011, at the University of California San Diego (UCSD), the Research Cyberinfrastructure (RCI) unit put out a call to faculty researchers to submit applications for participation in its Research Curation and Data Management pilot program. (RCI itself was formed following much planning, reporting, and organizing around service and infrastructure requirements in research data curation.) At UCSD eight pilot projects were approved - five emphasizing data curation needs, three steeped in storage needs. RCI is assisting the pilot projects "with the creation of metadata to make data discoverable and available for future re-use; with the ingest of data into the San Diego Supercomputer Center's (SDSC) new Cloud Storage system, which is accessible via high-speed networks; and with the movement of data into Chronopolis, a geographically-dispersed preservation system" (28).

Efforts such as the foregoing will establish leads to answers for a host of questions librarians have been considering, such as new staffing roles and new kinds of collaborations (not to mention a better understanding of whom to collaborate with and how to collaborate effectively). These efforts will also increase knowledge about organizational capacity for improving and expanding current infrastructure, likely leveraging it in unforeseen, innovative ways – an advancement important for both small and large institutions. As conveed below, a few libraries have developed service models worthwhile examining in depth, particularly for discerning common characteristics or practices that others might adopt and build on.

Examples of Data Management Services in Libraries

Following the NSF mandate - and, perhaps for some institutions, even preceding it - many libraries created new websites to convey information about the DMP requirement, what it meant to researchers, and how that particular library could help. An early exemplar was MIT's *Data Planning Checklist*, which posed questions that retrospectively mapped well to the five sections of the NSF's suggested approach to DMPs (*29*). Another web-based resource that developed just before the NSF requirement went into effect was the Association of Research Libraries' *Guide for Research Libraries: The NSF Data Sharing Policy*; it unpacks what a DMP is, states the leadership role that libraries harbor in this effort, offers guidance on how to help researchers craft a DMP, and aggregates a range of resources relevant to data curation and data management (*30*).

Since those heady first days of acting as early responders to the NSF mandate by assisting researchers with their questions and concerns, many libraries have arrived at service models, revising infrastructure or establishing it anew and, in some cases, creating new positions to support these promising service frameworks. There are too many to present in adequate detail here, but the examples highlighted below vary enough from each other to afford a picture of rich possibilities for other libraries to adapt for their local contexts.

Cross-campus collaborations in support of services for research data management mark one requisite for success. Besides distributing and sharing responsibility, collaborations between libraries and other entities at an institution help ensure that a diverse range of needs are investigated and met. Cornell University's Research Data Management Service Group (RDMSG, (*31*)), whose members are referred to as "consultants," brings together not only librarians and specialists in IT (such as security) but also people with experience in project management, copyright and intellectual property rights issues, high performance computing, and data management system design. The model for collaboration and consolidation endorsed by RDMSG resulted in part from extensive planning and gap analysis work, represented in their report, *Meeting Funders' Data Policies: Blueprint for a Research Data Management Service Group* (*32*). RDMSG has sponsorship from both the Senior Vice Provost for Research and the University Librarian; a faculty advisory board helps the group discern data management needs among faculty researchers, as well as determine additional resources and services for facilitating DMPs; and a management council exercises further oversight. The group is clear about the services it makes available, one of which is to "Provide a single point of contact that puts researchers in touch with specialized assistance as the need arises." In addition, it consolidates in a single directory the services, tools, and resources found across campus that are relevant to research data management, including tools for collaboration, guidance in intellectual property rights and data publication, and services in storage and backup, metadata creation, and data analysis and tools for collaboration.

Another necessity for data management services in libraries, especially as suggested by the University of Virginia model discussed above, is the training component, whether in the form of workshops, instruction sessions, or information sessions. The University of Minnesota (UMN) Libraries have crafted a suite of training opportunities to meet the needs of their faculty, students, and staff. Three librarians with expertise in the sciences and the social sciences form the foundation of support in the UMN Libraries for data management planning services. The workshop offerings range from "Creating a Data Management Plan for Your Grant Application," to "Introduction to Data Management for Scientists and Engineers," to "Data 101: Best Practices throughout the Data Life Cycle" (*33*). Perhaps most important, the workshop on creating a DMP meets UMN's requirements for continuing education in responsible conduct of research, thereby providing another incentive for faculty and students to enroll in it. At the UMN Libraries' website, online tutorials and workshops given by other departments and units on campus are also listed - these include sessions on technology training, intellectual property, quantitative data management, qualitative data management, and HIPAA data security training, better data searching, and help with grant proposal development (e.g., tools and resources for it, as well as guidance geared to graduate students seeking grant funding).

A chief requirement in NSF's DMP guidance is that data generated by a research project be preserved to ensure continued sharing and access. While quite a number of data repositories, or disciplinary repositories accepting data, exist, they tend to suit "big science" projects producing data at a far larger scale than many university-based researchers with NSF funding contend with, in reality. Libraries that run or manage institutional repositories (IRs) may be in a satisfactory position to accept "small science" data - although, as Salo asserts, most repository software applications are suited for *finished* scholarly publications such as journal articles and book chapters, rather than versions of data sets: "For data, permitting only the immutable is unacceptable [. . .] much of the value of data is precisely its mutability in the face of new evidence or new processes." (*18*).In addition, IRs that do accept data sets can usually accommodate raw data files but have few or no additional tools for data visualization (*18*).

Some institutions have started addressing the challenge of data deposits in a repository context. One is Purdue University, through its Purdue University Research Repository or PURR, and another is Rutgers University, through its RUresearch Data Portal, a part of its Rutgers University Community Repository, also known as RUcore. PURR is an instance of the Purdue-developed Hub Zero platform, devised for collaboration and for sharing of data and tools needed for research data in the sciences: "PURR provides workflows and tools for ingestion, identification and dissemination of data as well as services to ensure data security, fidelity, backup, and mirroring. Purdue Libraries will consult with investigators to facilitate selection and ingestion of data with the application of appropriate descriptive metadata and data standards as well as to provide appraisal of data for long-term preservation and stewardship" (*34*). RUresearch Data Portal allows a broad range of "research genres" that are defined largely by the type of data, or data container, germane to that particular genre (*35*). These include, but are not limited to, codebooks, experimental data, multivariate data, quantitative discrete data, and quantitative continuous data. The research domains represented in the RUresearch Data Portal, as of spring 2012, are cognitive science, computer science, environmental engineering, political science, and statistics. One of the service points in the RUresearch Data Portal is customization of a portal for searching and retrieval of one's data, made possible largely by a "sophisticated, flexible metadata strategy that can customize metadata to support your primary audience yet still be compatible with prevailing metadata standards" (*35*).

When it comes to supporting data management for grant-funded projects, some institutions are implementing cost-based models, particularly for storage and archiving services. The aforementioned PURR charges for extra project space (for the life of the project) and extra publication space (for ten years). To archive research data for a minimum five-year period, Johns Hopkins University charges a fee that is 2% of the direct total cost on an NSF grant, "with the option for an extension, and our expert support helping you prepare data for preservation and sharing" (*36*). Johns Hopkins is explicit about what its Data Archive offers - and thus what NSF PIs would be paying for, which is an archive that accepts discipline-agnostic data; a "data integration framework" enabling querying across the archived data; and a "preservation-ready system" (*36*).

Finally, a key component in data management service models is dedicated referencing of institutional policies and guidelines related to research activities, if not a distinct institutional policy for research data. In this particular service area, institutions in the U.K. are strides ahead of those in the U.S. The Digital Curation Centre, based in Edinburgh, Scotland, maintains a evolving online list of institutions with agreed-upon data management policies as well as a list of institutions that have completed policy drafts. The definition and implementation of an institution-wide policy for research data management has many dependencies, not least of which is obtaining buy-in from a spectrum of campus stakeholders. Such a commitment requires a common understanding among stakeholders (who, at the outset, may encompass librarians, IT staff, researchers, and administrators) of the issues, needs, and goals for management of research data throughout their life cycle. One place with an institutional policy regarding research data is Johns Hopkins. Its "Policy on Access and Retention of Research Data and Materials" (*36*) defines what is meant by research data and by the role of the "primary responsible investigator" and specifies how long the university will retain the data (for five years). Another institution with a research data policy is the University of Tennessee, which addresses responsibility, control, retention, and ownership of data, as well as the rights of the University to research data (*37*). Other considerations for an institutional policy on management of research data could include - but are not limited to - the following: a commitment to offer training and support opportunities, as well as guidelines and templates, to help researchers create DMPs; provision of tools and services for conducting preservation actions on data and retaining them to ensure their uninterrupted access; and a resolve not to relinquish rights to commercial entities to reuse or publish research data without making certain the institution continues to have the right to make the data publicly available (*38*).

Conclusion

The story of data management services in libraries reflects an abundance of still developing plot points and characters - ones that have begun to yield promising service frameworks, collaborations, tools, and new roles. An example of a tool that emerged a year after the May 2010 press release from the NSF is the DMP Tool, which walks researchers through a data management plan, allowing them to input the relevant information for each section, and then generates the plan (*39*). (It is up to the researcher to make sure the plan, after it is generated, does not exceed the two-page maximum length.) The tool also provides additional resources, such as DMP guidelines from not only the NSF but also the National Endowment for the Humanities and other funding agencies with similar requirements. The DMP Tool is a collaborative effort of seven institutions and organizations: California Digital Library, the Digital Curation Centre (U.K.), Smithsonian Institution, UCLA Library, University of Illinois at Urbana-Champaign, and University of Virginia Library. Developers of the tool are also enabling federated access to it for a growing number of universities and colleges.

Since the activation of the NSF requirement, several academic libraries have also created positions in which data curation or data management (or both) is the primary responsibility. Examples of some of these positions are "Data Services Librarian" (Kansas State University Libraries, (*40*)), "Data Curator" (Simon Fraser University Library,(*41*)), "Data Management Specialist" (Emory University Libraries, (*42*)), and "Manager, Data Management Services" (Johns Hopkins University Libraries, (*43*)). While these positions are based in an academic library, they are highly collaborative roles that work across library and, often, campus departments. They frequently require experience with project management and with curation of scientific data; knowledge of intellectual property rights issues, repository infrastructure, and metadata standards, as well as an awareness of rising trends, both locally and globally, in data curation and data management; excellent communication and interpersonal skills; and a dedication to providing the best possible patron service and support. In addition, in spring 2012, the Council on Library and Information Resources (CLIR), working with the Digital Library Federation (DLF), announced a new postdoctoral fellowship program in data curation, which the two organizations launched with funding from the Alfred P. Sloan Foundation. A main intent behind this effort is "to raise awareness and build capacity for sound data management practice throughout the academy" (*12*). In summer 2012, CLIR and DLF announced the first cohort of Data Curation Postdocoral Fellows.

The ultimate aim in telling the story of data as thoroughly and properly as possible, drawing on some of the paths and examples relayed above, is so that one's data will be able to be found, preserved, accessed, shared, used, and reused - not just in the current century but also beyond it. Additionally, it is so that researchers themselves will have a clear idea of where their own data are. As this chapter suggests, traction is gaining in favor of libraries, IT units, research administration, faculty, graduate students, and others working together to flesh out the story of data at their campuses. They are encouraging best practices and standards, reconfiguring roles and responsibilities strategically to meet demands in data management, and planning for infrastructure that aligns, rather than duplicates, services across an institution. In other words, the story of data in the 21st century and further is the story of a brave new world.

References

1. U.S. Office of Management and Budget. Federal Register Notice Re OMB Circular 110: Uniform Administrative Requirements for Grants and Agreements with Higher Education, Hospitals, and Other Non-Profit Institutions. http://www.whitehouse.gov/omb/fedreg_a110-finalnotice (accessed March 15, 2012).
2. Goldstein, S.; Oelker, S. Planning for Data Curation in the Small Liberal Arts College Environment. *Sci-Tech News* **2011**, *65*, 5−11. http://jdc.jefferson.edu/scitechnews/vol65/iss3/4/ (accessed March 2012).
3. Borgmann, C. L. Research Data: Who Will Share What, with Whom, When, and Why? China−North America Library Conference, Beijing. http://works.bepress.com/borgman/238/ (accessed April 2012).
4. Ogburn, J. The Imperative for Data Curation. *Portal: Libraries and the Academy* **2010**, *10*, 241−246. http://muse.jhu.edu/journals/pla/summary/v010/10.2.ogburn.html (accessed July 2012).
5. Marcum, D.; George, G. *The Data Deluge: Can Libraries Cope with E-Science?*; ABC-CLIO, 2009.
6. Witt, M.; Carlson, J.; Brandt, D. S.; Cragin, M. H. *Int. J. Digital Curation* **2009**, *4*, 93−103.
7. Delserone, L. M. At the Watershed: Preparing for Research Data Management and Stewardship at the University of Minnesota Libraries. *Library Trends* **2008**, *57*, 202−210. https://www.ideals.illinois.edu/handle/2142/10670 (accessed July 2012).
8. Gold, A. Cyberinfrastructure, Data, and Libraries, Part 1. A Cyberinfrastructure Primer for Librarians. *D-Lib Magazine* **2007**, *13*, np. http://www.dlib.org/dlib/september07/gold/09gold-pt1.html (accessed March 2012),
9. Gold, A. Cyberinfrastructure, Data, and Libraries, Part 2. Libraries and the Data ChallengeSRoles and Actions for Libraries. *D-Lib Magazine* **2007**, *13*, np. http://www.dlib.org/dlib/september07/gold/09gold-pt2.html (accessed March 2012).
10. Hay, T.; Trefethen, A.; The Data Deluge: An e-Science Perspective. In *Grid Computing: Making the Global Infrastructure a Reality*; Berman, F., Fox, G. C., Hey, A. J., Eds.; Wiley and Sons: New York, 2003; pp 809−824.
11. Salo, D. *Research Data Management across the Disciplines.* https://mywebspace.wisc.edu/dmrovner/web/LIS341.htm (accessed July 2012).

12. Council on Library and Information Resources. Awards & Fellowships: CLIR/DLF Data Curation Postdoctoral Fellowship. http://www.clir.org/fellowships/datacuration (accessed June 2012).
13. Hey, T., Tansley, S., Tolle, K., Eds.; *The Fourth Paradigm*; Microsoft Research: Redmond, WA, 2009.
14. Jasny et al. Introduction: Again, and Again, and Again . . . *Science* **2011**, *334*, 1225. http://www.sciencemag.org/content/334/6060/1225.full (accessed March 2012).
15. National Science Foundation. Scientists Seeking Funding Will Soon Be Required To Submit Data Management Plans, 2010. http://www.nsf.gov/news/news_summ.jsp?cntn_id=116928 (accessed March 2012).
16. Ramirez, M. L. Whose Role Is It Anyway? A Library Practitioner's Appraisal of the Digital Data Deluge. *Bull. Am. Soc. Inf. Sci. Technol.* **2011**, *37*, 21–23. http://www.asis.org/Bulletin/Jun-11/JunJul11_Ramirez.html (accessed July 2012).
17. Heidorn, P. B. The Emerging Role of Libraries in Data Curation and e-Science. *J. Library Admin.* **2011**, *51*, 662–672. http://www.tandfonline.com/doi/abs/10.1080/01930826.2011.601269 (accessed July 2012).
18. Salo, D. Retooling Libraries for the Data Challenge. *Ariadne.* **2010**, *64*. http://www.ariadne.ac.uk/issue64/salo (accessed March 2012).
19. Lyon, L. The Informatics Transform: Re-Engineering Libraries for the Data Decade. *Int. J. Digital Curation* **2012**, *7*, 126–138. http://www.ijdc.net/index.php/ijdc/article/view/210 (accessed July 2012).
20. Reznik-Zellen, R. et al. Tiers of Research Data Support Services. *J. eScience Librarianship* **2012**, *1*. http://escholarship.umassmed.edu/jeslib/vol1/iss1/5/ (accessed July 2012).
21. Data Asset Framework. http://www.dcc.ac.uk/resources/repository-audit-and-assessment/data-asset-framework (accessed March 2012).
22. Digital Curation Centre. UK Institutional Data Policies. http://www.dcc.ac.uk/resources/policy-and-legal/institutional-data-policies (accessed June 2012).
23. Parham, S. Testing the DAF for Implementation at Georgia Tech. Presented at the 6th International Digital Curation Conference, December 6–8, 2010, Chicago, Illinois. http://smartech.gatech.edu/handle/1853/39786 (accessed April 2012).
24. Peters, C.; Dryden, A. R. Assessing the Academic Library's Role in Campus-Wide Research Data Management: A First Step at the University of Houston. *Sci. Technol. Libr..* **2011**, *30*, 387–403. http://www.tandfonline.com/doi/full/10.1080/0194262X.2011.626340 (accessed March 2012).
25. University of Massachusetts at Amherst. e-Science Portal for New England Librarians. http://esciencelibrary.umassmed.edu/ (accessed April 10, 2012).
26. Creamer, A.; Morales, M.; Crespo, J.; Kafel, D.; Martin, E. An Assessment of Needed Competencies To Promote the Data Curation and Management Librarianship of Health Sciences and Science and Technology Librarians in New England. *J. eScience Librarianship*, 2012. http://escholarship.umassmed.edu/jeslib/vol1/iss1/4/ (accessed July 2012).

27. Sallans, A.; Lake, S. How to Re-Tool Librarians for Data Curation. Presented at the 6th International Digital Curation Conference, December 6–8 2010, Chicago, Illinois. http://www.dcc.ac.uk/webfm_send/297 (accessed September 2012).
28. University of California San Diego. Research Cyberinfrastructure. http://rci.ucsd.edu/pilots/index.html (accessed May 2012).
29. Massachusetts Institute of Technology. Data Management and Publishing: A Data Management Checklist. http://libraries.mit.edu/guides/subjects/data-management/checklist.html (accessed May 2012).
30. Association of Research Libraries. Guide for Research Libraries: The NSF Data Sharing Policy. http://www.arl.org/rtl/eresearch/escien/nsf/index.shtml (accessed May 2012).
31. Research Data Management Service Group. About the RDMSG. https://confluence.cornell.edu/display/rdmsgweb/About (accessed May 2012).
32. Block, B.; et al. Meeting Funders' Data Policies: Blueprint for a Research Data Management Service Group. http://ecommons.library.cornell.edu/handle/1813/28570 (accessed May 2012).
33. University of Minnesota Libraries. Data Management: Workshop and Training. https://www.lib.umn.edu/datamanagement/workshops (accessed June 2, 2012).
34. Purdue University Research Repository. https://research.hub.purdue.edu/ (accessed June 2012).
35. Rutgers University. RUcore: RUresearch Data Portal. http://rucore.libraries.rutgers.edu/research/ (accessed June 7, 2012).
36. Johns Hopkins University. Data Management Services. http://dmp.data.jhu.edu/ (accessed June 2012).
37. University of Tennessee, Office of Research. Policies: Administrative Policies. http://research.utk.edu/forms_docs/policy_research-data.pdf (accessed June 7, 2012).
38. Pryor, G.; Donnelly, M. Skilling Up To Do Data: Whose Role, Whose Responsibility, Whose Career? *Int. J. Digital Curation* **2009**, *4*, 158–170. http://www.ijdc.net/index.php/ijdc/article/view/126 (accessed July 2012).
39. DMP Tool: Guidance and Resources for Your Data Management Plan. https://dmp.cdlib.org/ (accessed July 2012).
40. K-State Libraries. Employment at K-State Libraries: Data Services Librarian. http://ksulib.typepad.com/jobs/2011/06/data-services-librarian.html (accessed June 2012).
41. Taylor, L. N. *Laurie N. Taylor (blog)*. Job: Data Curator at Simon Fraser University Library, 2012. http://laurientaylor.org/2012/02/04/job-data-curator-at-simon-fraser-university-library/ (accessed June 10, 2012).
42. Emory University Libraries. Data Management Specialist. http://web.library.emory.edu/sites/web.library.emory.edu/files/edc-data_management_specialist_pa_120811.pdf (accessed June 2012).
43. Johns Hopkins University. Human Resources: Data Management Specialist (Manager, Data Management Services). https://hrnt.jhu.edu/jhujobs/job_view.cfm?view_req_id=52688&view=sch (accessed June 2012).

Chapter 8

Research Data Management and the Role of Libraries

Mary C. Schlembach[*,1] and Carol A. Brach[2]

[1]Engineering, Physics, and Astronomy Librarian,
University of Illinois at Urbana-Champaign, Urbana, Illinois 61801
[2]Engineering Librarian, University of Notre Dame,
Notre Dame, Indiana 46556
*E-mail: schlemba@illinois.edu

>Academic research libraries in the US and abroad are already playing roles as leaders in areas where libraries and librarians can bring significant value to data management efforts. With new data management stewardship mandates by national government agencies in place, libraries need to take advantage of new opportunities in data stewardship and curation. Focusing on e-science and the management of scientific data, this chapter highlights many of the data management programs developed at academic libraries.

Introduction

While data has been collected, stored, preserved, lost and repurposed for centuries (particularly in the social sciences), data science—as both an academic discipline and as a library service—has arrived. Everywhere people are talking about data and libraries are no exception. In fact, academic research libraries in the US and abroad are already playing roles as leaders in areas where libraries and librarians can bring significant value to data management efforts. Places in the data management life cycle where libraries have been building expertise are institutional repositories, standards for description of data and research objects, accuracy, and accessibility.

The National Institute of Health's requirement for data plans for grants over $500K and the National Science Foundations' requirement (*1*) for a half-page to two-page data management plan (DMP) contains numerous elements that require careful consideration, many of which have not traditionally been part of researchers' normal activities. Many researchers will have questions or need advice about where and how the data will reside. Many also consider collected data to be "their" data to do with as they want or need. This, in turn, creates many ethical questions and circumstances. What are the issues surrounding privacy? How can data be made accessible and how should it be described using standard conventions such as metadata to enable data sharing? Where are the viable and cost effective repositories to store and share data? Who has access to the data and at what point in time? When do the others start to have access to the data? What happens if the data is misused?

Data for all disciplines is growing exponentially both in size and formats and is becoming more intensive and collaborative. Thus it is becoming more important to share data even across disciplines where barriers formerly existed and increasingly researchers and scholars are actively looking for ways to reuse data.

The Association of Research Libraries (ARL) report *New Roles for New Times: Digital Curation for Preservation* outlines new roles for librarians and strategies for collaboration in research libraries (*2*). The National Academy of Sciences has made several recommendations for data standards which emphasizes the role of the institutions where research is generated (*3*). Heidorn points out that research libraries are best suited to lead the way for data curation and preservation (*4*). Newton, Miller and Bracke expand on the roles of librarians in data stewardship: data identification, mediation, selection and appraisal, and preparation – all of which can be integrated into research data and compilation (*5*). New roles for librarians are taking hold as they take a lead role in establishing institutional repositories, developing standards for metadata, and teaching other data management assistance skills. There is a new emphasis with librarians and libraries moving to support faculty digital publishing activities. As these roles mature, the library's relevance to the faculty—and, consequently, the institution overall—will increase" (*4, 5*).

There are still many questions that need to be answered about who are the data specialists in our libraries whose expertise could be leveraged for purposes of both "inreach" (librarian colleagues in data management concepts and practice) and outreach (getting the word out to faculty researchers that the library is ready to help). Given the increasing emphasis on the ability to understand and work with data, as well as to manage it, it becomes incumbent on librarians and faculty to work together to educate students early – to impart consistent advice on how to "do" data planning (*7*).

Studies done by Tenopir, et al. found that there is a need for projects to build the infrastructure, policies and best practices to address data sharing and curation. There are some tools, but often there is little knowledge or satisfaction with metadata tools. This is where librarians and libraries can assist researchers to help them prepare data properly and make it both retrievable and reusable. Importantly, libraries need to create standards and systems so researchers can easily submit their data using flexible and efficient metadata schemas. Libraries need to address the culture of data sharing, preservation, use and scaling to other disciplines. It is building this infrastructure that helps science move forward (*8*).

Over the past few years, there have been several research initiatives around the world studying library and librarian roles and skill sets as they change to support research and scholars. Many of these studies have focused on how researchers are archiving (or not) and sharing (or not) their research data. European organizations such as the Joint Information Systems Committee (*9*), the Association of European Research Libraries (*10*), Permanent Access to the Records of Science in Europe (PARSE.Insight) (*11*), Research Libraries UK (*12*), and the Research Information Network (*13*) with the British Library have explored researchers' information needs, the preservation of research data, and the rapid changes in research that have impacted researchers, librarians, and other support services of data management. Data management is undergoing rapid and remarkable changes and with these worldwide initiatives libraries have the opportunity to be a major player.

Discipline Repositories

For the past few years, even before the NIH and NSF mandated data management plans, government agencies and scientific societies --often in collaboration with libraries-- have been coordinating efforts for disciplinary repositories such as PubMed, Dryad, and arXiv. Discipline repositories have been successful with uptake from researchers in their respective fields since many of them have been involved in the development. Sustainability of discipline repositories has become more of an issue, however, and repositories such as arXiv have had to request voluntary donations (in the form of "institutional memberships") to maintain levels of service and to continue their growth.

The Open Access Directory at Simmons College maintains a list of repositories based on discipline at http://oad.simmons.edu/oadwiki/ Disciplinary_repositories (*14*).

Institutional Repositories

For those disciplines which may not currently have a data repository or may be more restrictive in what types of data are accepted, institutional repositories have become a resource crucial for the success of data management plans. Hundreds of libraries have begun institutional repositories over the past 15 years. Starting as a method for researchers to provide their research in an open environment, they have progressed to a storehouse for electronic theses and dissertations and data depository. Institutional repositories have now developed into a new medium in which to archive institutional researchers' data and store the intellectual output to fulfill granting agencies requirements. This, in turn, also provides a broader impact for the researcher, their institution, and the funding agency.

The goals of institutional repositories are to provide easy access, increase the visibility of scholarship and data, and preserve and archive the institution's research content, particularly content that has been born digital.

As Crow (2002) states, organizing and maintaining digital content—as well as supporting faculty as information contributors and end users—should remain the responsibility of the library. Libraries are best-suited to providing much of the document preparation expertise (document format control, archival standards, etc.) to help authors contribute their research to the institution's repository. Similarly, libraries can most effectively provide much of the expertise in terms of metadata tagging, authority control, and the other content management requirements that increase access to, and the usability of, the data itself (*6*).

Major Players

Several institutions are already major players and have assumed leadership roles that can guide other libraries in moving forward with data management services on their campuses.

Johns Hopkins University

Started in 2011, the Sheridan Libraries at Johns Hopkins University provide guidance services for researchers with an emphasis on data management plans for National Science Foundation proposals. The Hopkins Data Management Services (DMS) (*15*) has consultants which help researchers assess and describe the data produced, and review draft data management plans. However, work began on establishing an infrastructure well before the launch (*16*).

JHU's team of consultants consists of a manager and two librarians. The group provides guidance on writing data management plans and provides an archive for research data. They also provide workshops so that PIs can learn what content should be included in data management plans, or learn about JHU policies on data management and other data-related services (*17*).

Prior to the launching of the DMS group's website, in 2008 Johns Hopkins University established policies that address other important issues such as data ownership and retention, policies about work that involves human subjects, the protection of intellectual property, and additional policies that address issues related to data management.

The DMS website explains in detail, how the DMS group can help, including a very important reason for choosing to use DMS services: They provide for researchers "a high-quality, tailored data management plan specific to your research and the NSF Directorate may improve the competitiveness of your proposal, particularly as NSF moves to a more systematic review and implementation of data management plans" (*16*).

Under "guidance" they lay out their feedback process and show demonstrate how draft plans are handled. They also can provide examples written by other JHU researchers. They also offer to visit research groups and offer to plan other events to discuss their services in more detail.

A plan generally consists of a description of data that will be produced, how it will be managed during and afterwards, and how it will be shared. Arriving at a solution about how data will be shared and what data can be shared is another function of the DMS team. They stress that sharing data enhances the visibility of research, promotes collaboration and community-building.

DMS can help to formalize the data citation effort by properly crediting researchers, assigning a DOI, and using descriptive elements that are emerging as guidelines such as: title, author, data, distributor, versions and locator/identifier or release date and resource type. The DMS consultant can offer guidance to a PI about how to prepare a budget for data management and where in the budget the expense should be documented. Annual reporting and following up with a final project report must include any updates to the DMP with such information as where the data sets are deposited and how they are archived.

The final elements that are important for researchers, such as: finding and using commercial and publicly-available datasets and helping with the software that is needed to utilize the data; server management and hosting; and data storage including backing up and securing data are all addressed in the "Campus Resources" section of the website.

Massachusetts Institute of Technology

MIT has published a "Data Management and Publishing" subject guide that is loaded with links that can help Principal Investigators and grant writers to quickly locate information on an issue in question or find contact information for advice (*18*).

One of the main goals of the requirement for a data management plan is to give other researchers the ability to locate and use data that has been generated by NIH, NSF, USDA, or other grant funding. This is a simple idea, and from the vantage point of data stewardship, a commendable concept; but one that poses many questions for grant writers, libraries, and institutions. A short, up to two page report must contain numerous elements that require careful consideration, many of which lie outside the normal activities of researchers. There are also data life cycles to consider, how to describe the data, how much data will be generated, and the questions surrounding file formats. What software is needed to re-create the data? How long the data should be retained? Those are the *simple* questions. Some of the harder questions for researchers could be where will the data reside until it is made publicly available? What if there are privacy concerns? Are there patent and commercialization factors? How can I make the data accessible? How should it be described using standard conventions such as metadata to enable data sharing?

The MIT guide covers every conceivable question, but gives no answer to the question, "Who in the research group will be responsible for data management?" One possible way for researchers to respond to this question is to include a metadata specialist in the grant application process and have that person generate a budget for data description, accessibility, and storage. This is where the question of the role of the library comes into play. The funding agencies, in the interest of supporting the data management requirement, are allowing Principal Investigators (PIs) to include requests for funding to cover data management costs in the grant proposal budget.

The "Data Management and Publishing" guide links to a wealth of important information for grant writers that could be displayed in other ways. But there are many issues connected with data management, and being able to pick out one or more issues and read about them at when time permits is a real challenge. The process can seem overwhelming to some researchers who have little or no experience with planning for data management. Having input from a faculty member who has successfully navigated the process on the guide is reassuring. From the standpoint of being able to quickly add new information or update descriptions of numerous elements, displaying information in a subject guide format offers many advantages.

Purdue University

As early as 2006, Purdue University Libraries enlisted graduate students funded by university research centers to provide metadata and web-based software for their data management. This expanded into librarians working directly with scientists, campus IT specialists, faculty of various research intensive disciplines, and University administrators to work collaboratively investigating problems and solutions for research data management, preservation and access and resulted in the Distributed Data Curation Center (D2C2). D2C2offers several online tools for data management (*19, 20*).

Data Curation Profiles (http://datacurationprofiles.org/)

Purdue's Data Curation Profiles started with an IMLS grant awarded to Purdue University and the University of Illinois I-school. It is a survey formatted for subject librarians to initiate conversations with researchers about their data curation requirements. The accompanying Data Toolkit enables librarians and researchers to review data to assess the best practices and help identify collaborative data services.

Data Management Plan Self-Assessment Tool (https://research.hub.purdue.edu/resources/7)

This tool was developed as a result of the Data Curation Profiles. It can assist researchers' with their interpretation of the data into a data management plan.

Databib(http://databib.org/)

Similar to the Open Access Directory of Repositories at Simmons College, Databib is a descriptive and searchable list of data repositories which is useful for researchers who are not sure of the most appropriate repository for their data (*21*).

Purdue University Research Repository (PURR)
(https://research.hub.purdue.edu/)

In collaboration with the Office of the Vice President for Research, the D2C2 started to review NSF proposals and began analyzing data management plan success using the HUB Zero platform (HUB Zero is a NSF funded Web-based software package to provide virtual communities with collaborative research environments.) PURR is an online, collaborative data sharing platform to promote the data management needs of Purdue researchers. Researchers from other institutions can also participate, as long as at least one Purdue employee is involved. When a project is begun, subject librarians are notified and are given the opportunity to consult with the project manager(s) on their data management needs. PURR provides a data management plan template with information regarding its general use, data object identifier (DOI) standards, and commitment towards being a Trustworthy Digital Repository. The PURR template has been identified in 34% of submitted NSF proposals (*22, 23*).

The primary goals of PURR are to develop knowledge of the strengths and weaknesses of distributed systems data; develop collaborative librarian – researcher projects; develop strategies for the access of disciplinary data tools.

One of the greatest strengths of PURR is that the datasets are assigned Digital Object Identifiers (DOIs) to expand the impact and accessibility to researchers. DOIs are a recognized standard for all electronic scientific literature.

University of Virginia

The University of Virginia Library has been providing data management support for all funding agencies, including NSF, since May 2010 when the NSF Data Management Plan requirement was announced. The Scientific Data Consulting Group (SciDaC) located in the Charles L. Brown Science & Engineering Library consults with researchers on managing their data throughout the entire data life cycle. Also provided by this group is a checklist for using the DMP tool which includes where to send a copy of the plan so that it can be reviewed by the SciDaC group (*24*).

The Virginia SciDaC group, in partnership with the California Digital Library, developed the DMP Tool http://dmp.cdlib.org, an online tool to help researchers generate data management plans. Any institution and researcher can use the DMP Tool. Virginia also uses the DMP Tool Guide on their website (*25*).

The University of Virginia has an entire suite of webpages with helpful information for researchers to view in order to get a better understanding about each step of the data management planning process. They include information on the research life cycle and emphasize the benefits of good data management practices. Types of data and advice about choosing the best file formats for data sharing are covered. File version control and tracking changes if research involves more than one person, as well as security, storage and backups round out the sections on organizing data.

The section on funding guidelines explicitly includes the main agency guidelines (NSF, NIH, and NASA) and offers help for others. The "Why and

What Data Should be Documented" section gives examples and links to the webpage for properly citing data.

Finally, there is a web page that gives important information about archiving and data sharing requirements for funding agencies and for some journals such as those of the Nature Publishing Group. This section includes a link to other information including the University of Virginia Data Retention Policy and links to ownership and privacy information. The University of Virginia website, like some others, also gives links to Community-endorsed Repositories and other Subject Specific Repositories.

University of North Carolina

Similar to Purdue, the University of North Carolina at Chapel Hill established a task force to conduct an environmental scan of digital research and data stewardship, including policies and trends, issues and types of data currently being collected, and developing principles and actions for the future (*26*). Their comprehensive process included collecting Research Data Profiles from a sampling of other institutions. Some of their environmental scan data is "re-used" in Table 1 where it is combined with a survey of the Engineering Libraries Division (ELD) members of the American Society of Engineering Education (ASEE) to give a snapshot of how institutions are addressing data management requirements.

Guided by underlying assumptions that scholarship is inherently collaborative, the UNC guidelines show that university policies must be flexible. One reason is that there is an enormous diversity of types of data and so a "one-size-fits-all" policy would not be appropriate. The task force went on to determine which funding agencies and publishers require deposition of data into repositories and their Table 1 gives examples. So, besides the NSF and NIH, the Department of Energy (DOE), National Aeronautics and Space Administration (NASA), and the National Oceanic and Atmospheric Administration (NOAA) agencies require data stewardship along with the American Association for the Advancement of Science (AAAS) and Nature journals.

UNC found that, as might be imagined, most data policies address intellectual property issues, and fewer mention data sharing or how to comply with policies. They surveyed researchers at UNC and the report contains information from humanities and social sciences as well as science-related disciplines. A major take-away for libraries is that UNC researchers are not aware of library services and standards. Funding for long term data storage by departments is not adequate and that the sharing of research data adds another dimension to research activity that requires additional time. And finally, the expectation should be communicated from the highest levels of the university that provisions for data management and sharing are required elements of grant applications because such elements enhance the university's stature and reputation. Detailed recommendations that have policy and infrastructure implications are included that can be models that other libraries can use to begin organizing their efforts to address data stewardship.

Table 1. University Plans

Institution	Data Management Plans	Data/Research Policies
Arizona State University	http://researchadmin.asu.edu/dmp	http://www.asu.edu/aad/manuals/rsp/rsp604.html
California Digital Library	https://dmp.cdlib.org/	http://www.cdlib.org/about/policies.html
Case Western Reserve University	http://case.edu/its/researchcomputing/datamanagement/	http://www.case.edu/its/policies.html
Columbia University	http://researchinitiatives.columbia.edu/nsf-data-management-plan	http://www.columbia.edu/cu/compliance/docs/data_management/
Duke University	http://library.duke.edu/data/guides/data-management/	http://www.provost.duke.edu/pdfs/fhb/FHB_App_P.pdf
Johns Hopkins University	http://dmp.data.jhu.edu/	http://dmp.data.jhu.edu/policies-and-resources/jhu-policies/
Massachusettes Institute of Technology	http://libraries.mit.edu/guides/subjects/data-management/nsf-dm-plan.html	http://web.mit.edu/policies/14/14.1.html
Ohio State University	http://orrp.osu.edu/irb/training/rcr/nsf.cfm	http://ocio.osu.edu/policy/policies/policy-on-institutional-data/
Purdue University Libraries	http://www.lib.purdue.edu/scholarly/data.html	http://research.hub.purdue.edu/content/article/51
Stanford University	restricted to SULAIR	http://rph.stanford.edu/2-10.html
Tufts University	http://researchguides.library.tufts.edu/datamanagement	http://researchadmin.tufts.edu/?pid=64&c=17
University of Arkansas	http://vpred.uark.edu/NSF-Data-Management-Plan.pdf	http://vcfa.uark.edu/Documents/3095.pdf
University of Idaho	http://www.uidaho.edu/research/fundingagencies/proposal/nsfdatamanagementplan	http://www.uidaho.edu/cnr/taylor/research/data/datamanagementpolicy

Continued on next page.

Table 1. (Continued). University Plans

Institution	Data Management Plans	Data/Research Policies
University of Illinois at Urbana Champaign	http://search.grainger.uiuc.edu/top/NSF_DMP_template.pdf	http://www.ideals.illinois.edu/handle/2142/5
University of Kentucky	http://www.research.uky.edu/pdo/	http://www.rgs.uky.edu/ori/data.htm
University of Massachusettes Amherst	http://www.umass.edu/research/policy-procedure/nsf-data-management-plan	http://www.massachusetts.edu/policy/datacomputing-policies.html
University of Michigan	http://www.lib.umich.edu/research-data-management-and-publishing-support/nsf-data-management-plans	http://spg.umich.edu/pdf/601.12.pdf
University of New Hampshire	http://libraryguides.unh.edu/content.php?pid=250661&sid=2567227	http://www.usnh.edu/olpm/UNH/VIII.Res/?C.htm
University of North Carolina	http://www.lib.unc.edu/datamanagement/index.html	http://www.nd.edu/~cszambel/UNCData%20Governance%20Policy%20(1).pdf
University of Notre Dame	restricted to ND.EDU	http://or.nd.edu/policies-and-procedures/procedures/data-retention-and-access/
University of Pittsburgh	http://www.pitt.edu/~offres/policies/NSF-DMP-Examples.pdf	http://www.provost.pitt.edu/documents/RDM_Guidelines.pdf
University of Rochester	http://www.rochester.edu/ORPA/ORPA-L/orpaL2010/NSF_proposalAwardGuide.htm	http://www.rochester.edu/ORPA/policies/retent.pdf
University of Utah	http://www.science.utah.edu/research/materials/data-mgmt-plan-desc.pdf	http://www.regulations.utah.edu/it/4-001.html
University of Virginia	http://www2.lib.virginia.edu/brown/data/	http://www2.lib.virginia.edu/brown/data/datarights.html

Institution	Data Management Plans	Data/Research Policies
University of Washington	http://escience.washington.edu/blog/writing-nsf-data-management-plan	http://www.washington.edu/uwit/im/dmc/docs/UWDataManagemenPolicyV19.pdf
Virginia Commonwealth University	http://www.research.vcu.edu/vpr/resources/grant_proposal.htm	http://www.research.vcu.edu/p_and_g/pdf/FNL%20Data%20Ownership,%20Retention,%20Access%20%20BOV%205-09.pdf
Virginia Tech	http://www.research.vt.edu/proposal-development-resources/announcements/2011/helpful-links-developing-nsf-data-management-plans.php	http://www.policies.vt.edu/7100.pdf

e-Science Portal for New England Librarians

The e-Science Portal was started by the University of Massachusetts Medical School with funding from the National Network of Libraries of Medicine. Content is provided by New England research librarians primarily in health, biological and the physical sciences. The main goals are to teach and discuss e-Science and the impact on librarianship and the discliplines that In Spring 2012 the focus was expanded to other national e-science data management projects and organizations (27).

The site provides access to several data curation, data citation, metadata, and data tools web sites and publications. As part of the librarian educational goals of the portal, there are links to continuing education programs, I-schools with data courses, and compentency skills and resources for librarians. It also hosts programs for e-Science librarians and posts announcements for related meetings and seminars.

University of Illinois at Urbana-Champaign (UIUC)

The Grainger Engineering Library Information Center at UIUC developed a set of documents and presentations describing the NSF Data Management Plan (DMP) requirement, with particular emphasis on the requirements of the NSF Engineering Directorate. Among the documents prepared by Grainger Library staff was an NSF grant Data Management Plan template designed to assist grant preparers in the preparation of the DMP component of their proposal (28). This DMP template was described in presentations that Grainger librarians made to College of Engineering departmental and Center IT staff, business managers, department heads, research officers, and other faculty and staff. In the course of these discussions, it became clear that various individuals, in addition to the investigators themselves, were at least partially responsible for writing the DMP sections of the grant proposals. The template was made available via the Grainger Library web site.

Beginning in July 2011, the UIUC Library was given permission to examine the campus's NSF proposals as they were submitted within the NSF Fastlane system. The Library conducted a preliminary analysis of the NSF proposals submitted between July 2011 and March 2012. A total of 712 NSF grants submitted to the NSF Fastlane system were examined. Of these documents, 465 were considered "valid" proposals for the purpose of evaluating their Data Management Plans. There were 68 proposals that used the Grainger Library DMP Template.

DMP Tool

The Data Management Planning Tool "DMP" tool was started by the California Digital Library and has been widely adopted by libraries seeking to establish themselves as campus partners in addressing the new data management requirements (24). All researchers are welcome to use the DMP Tool, but those at member institutions benefit from having links to resources and services available at their respective institutions. The tool helps enable researchers with step-by-step instructions to create data management plans for specific funding agencies. It is available at https://dmp.cdlib.org/.

The DMP tool provides a template to guide researchers applying for NSF and NIH grants to include the following major topics in their grant applications:

- What data will be generated and or used from another study?
- What data and metadata standards will you be employing so that others can understand and re-use your data?
- How are you handling data storage and back-up during your project?
- How are you ensuring long-term access to your data after the project is complete?
- Which research are you sharing and how are you disseminating the data?
- Policies for access and sharing including why data will not be shared and policies on data re-use.

Challenges for Libraries

The biggest challenge for libraries involved in data curation activities is uptake by their institutions and researchers. Government mandates do help but many researchers do not want to submit their data until they have had a chance to publish.

Another challenge is the learning curve in data stewardship which includes devoting time to learning new technologies or methodologies of data curation. It is also a fact that both time and computing resources are necessary to provide data management services and it is a large investment – in terms of people and hardware for a library to make. Becoming a crucial and trusted part of a researcher's lab takes time and effort. And if a researcher does not see the need for assistance until the end of a project, or believes the involvement of the library is risky to the integrity of the publication process, then librarians have an even more challenging role to play. However, as Heidorn points out, if libraries do not accomplish this task, others will step up to take it on (4). This value of collaboration between libraries and researchers cannot be overestimated. Researchers potentially risk future grants if they do not have concrete data management plans and libraries risk being left out of the research agendas of their respective institutions. The traditional library now needs to become more agile and creative.

For those institutions or disciplines without repositories, another challenge is where to store the data. This, in turn, leads to questions about those schools or disciplines having less opportunity for federal grant funding because of the lack of an infrastructure which is expensive to start up. The National Institutes of Health have provided the PubMed repository. However, many of the National Science Foundation's Directorates do not have discipline repositories to support smaller institutions' data curation.

Resources for Librarians

ARL Data Sharing Support Group on Google Groups, ARL Guide for Research Libraries: The NSF Data Sharing Policy http://www.arl.org/rtl/eresearch/escien/nsf/index.shtml

The Data Conservancy program is part of the Johns Hopkins Sheridan Library which focuses on a data curation and archiving. Sayeed Choudhury, project principal investigator and associate dean for Library Digital Programs at Johns Hopkins University began the project by procuring an NSF's Office of Cyberinfrastructure grant. http://www.nsf.gov/dir/index.jsp?org=OCI

Choudhury outlined the work of the Data Conservancy to the Library of Congress on June 7, 2010. http://www.digitalpreservation.gov/news/2010/20100623news_article_Data_Conservancy.html

More important to libraries who are launching fledgling data management repositories, to learn more about the Data Conservancy software there is a blueprint related to the set-up of an instance that may be of interest and can be found at http://dataconservancy.org/community/blueprint/ (*29*)

Digital Preservation Coalition works to preserve digital resources in the UK and work with others for best practices. http://www.dpconline.org/

Conclusion

No one will have all the required skills. Different data types and disciplines will develop new standards and best practices and libraries must keep up with the developments. To best accomplish this, the subject specialists or data curators are best working in collaboration with researchers as the data is being processed and utilized. This helps ensure that data is used accurately and the metadata schema best represents the data.

Libraries need to be a part of the new data trends. It is part of our heritage and as more and more traditional libraries become a thing of the past, it is also the future of libraries.

References

1. National Science Foundation Dissemination in Sharing of Research Results. http://www.nsf.gov/bfa/dias/policy/dmp.jsp (accessed March 30, 2012).
2. Walters, T.; Skinner, K. New Roles for New Times: Digital Curation for Preservation. *Association of Research Libraries Report*, March, 2011.
3. Committee on Ensuring the Utility and Integrity of Research Data in a Digital Age; National Academy of Sciences In *Ensuring the Integrity, Accessibility, and Stewardship of Research Data in the Digital Age*; National Academies Press: Washington, DC, 2009.
4. Heidorn, P. B. The Emerging Role of Libraries in Data Curation and E-science. *Journal of Library Administration* **2011**, *51*, 662–672.
5. Newton, M. R.; Miller, C. C.; Bracke, M. S. Librarian roles in institutional repository data set collecting: Outcomes of a research library task force. *Collection Manage.* **2011**, *36* (1), 53–67.

6. Crow, R. The Case for Institutional Repositories: A SPARC Position Paper. *ARL Bimonthly Report 223*, 2002
7. Hswe, P.; Holt, A. Joining in the Enterprise of Response in the Wake of the NSF Data Management Planning Requirement*Research Library Issues: A Bimonthly Report from ARL,CNI, and SPARC*, February, 2011, pp 11–17.
8. Tenopir, C.; Allard, S.; Douglass, K.; Aydinoglu, A. U.; Wu, L.; Read, E.; Manoff, M.; Frame, M. Data sharing by scientists: Practices and perceptions. *PLoS ONE* **2011**, *6*, e21102.
9. Joint Information Systems Committee. http://www.jisc.ac.uk/ (accessed April 2, 2012).
10. LIBER Association of European Research Libraries (LIBER). http://www.libereurope.eu/ (accessed April 2, 2012).
11. PARSE.Insight. www.parse-insight.eu (accessed March 15, 2012).
12. Research Libraries UK. http://www.rluk.ac.uk/ (accessed April 2, 2012).
13. Research Information Network Research Information Network. http://www.researchinfonet.org/ (accessed April 20, 2012).
14. Simmons College Open Access Directory at Simmons College. http://oad.simmons.edu/oadwiki/Disciplinary_repositories (accessed April 18, 2012).
15. Johns Hopkins Library Data Management Services. http://dmp.data.jhu.edu/about-us/data-management-services-team/ (accessed April 20, 2012).
16. Johns Hopkins Library DMP Tools. http://dmp.data.jhu.edu/assistance/guidance-with-preparing-your-nsf-data-management-plan/ (accessed April 20, 2012).
17. Johns Hopkins Library Overview of Plans. http://dmp.data.jhu.edu/plan-basics/overview-of-plan-requirements/ (accessed April 16, 2012).
18. MIT Library Data Management Checklist. http://libraries.mit.edu/guides/subjects/data-management/checklist.html (accessed April 17, 2012).
19. Purdue University Distributed Data Curation Center. http://d2c2.lib.purdue.edu/ (accessed April 18, 2012).
20. Witt, M.; Carlson, J. R.; Cragin, M. R.; Brandt, D. S. Constructing Data Curation Profiles. *Int. J. Digital Curation* **2009**, *4*.
21. Databib. http://databib.org/# (accessed April 20, 2012).
22. Purdue University Research Repository (PURR). http://research.hub.purdue.edu/ (accessed May 18, 2012).
23. Witt, M. Co-designing, co-developing, and co-implementing an institutional data repository service. *J. Libr. Admin.* **2012**, *52*, 172–188.
24. University of Virginia Library Scientific Data Consulting. http://www2.lib.virginia.edu/brown/data/DMP_Support.html (accessed April 27, 2012).
25. DMPTool. https://dmp.cdlib.org/help/guide (accessed April 28, 2012).
26. University of North Carolina Research Data Stewardship. http://sils.unc.edu/sites/default/files/general/research/UNC_Research_Data_Stewardship_Report.pdf (accessed May 14, 2012).
27. e-Science Portal for New England Librarians. http://esciencelibrary.umassmed.edu/index (accessed July 25, 2012).

28. Grainger Engineering Library. http://search.grainger.uiuc.edu/top/ (accessed April 29, 2012).
29. Choudhury, S. Integration of Digital Library Services. http://www.library.jhu.edu/departments/librarydean/integration.html (accessed May 14, 2012).

Chapter 9

Preparing To Support Research Data Sharing

Ye Li* and Lori Tschirhart

Shapiro Science Library, University of Michigan,
Ann Arbor, Michigan 48109
*E-mail: liye@umich.edu

> When national funding agencies introduced data management requirements for grant proposals, details were scant and researchers turned to research-supporting staff for assistance with compliance. Support staff has experienced a sudden demand for e-Science and data sharing knowledge and expertise, often before institutional infrastructures and strategic plans have been developed. As a part of the research-supporting system, we share our learning paths, resources, and strategies here. We also describe and analyze emerging needs in the Chemistry domain to demonstrate discipline-specific data sharing issues and approaches used to customize services for local research communities.

Introduction

When we started our current positions as subject specialists at the University of Michigan's Shapiro Science Library in 2009, we had not anticipated how quickly e-Science, data sharing and data management would become central aspects of our job. What started with the 2003 Atkins Report by the National Science Foundation (NAF) Blue-Ribbon Advisory Panel on Cyberinfrastructure, which declared the need and potential for an e-Science revolution for science and engineering research in the U.S. (*1*), ended with funding agencies including and mandating data management components to their guidelines. In between, there were several important articles published on the subject (e.g., Anna K. Gold's two-part article (*2, 3*)), as well as a 2009 guide discussing how subject specialists could collaborate with researchers on e-Science projects at Purdue University (*4*). Now there are many other reports from NSF, other national agencies, and related organizations that provide an overview of cyberinfrastructure and e-Science in the

© 2012 American Chemical Society

U.S. and around the world. Some selected documents are linked on our research guides at http://guides.lib.umich.edu/ci. The reports clarify the big picture and sometimes present domain-specific research needs and challenges.

However, when funding agencies were only just beginning to include and mandate data management components in their guidelines, paths to meeting the guidelines were not always obvious. The burden was even more challenging for fields like Chemistry since it is considered a "small science," a "long-tail science," and a somewhat proprietary discipline without a data sharing tradition (5). A recent publication, *The fourth Paradigm: data-intensive Scientific Discovery* (6), highlights the research potential for those domains not yet obvious part of "big data." For example, chemistry as a basic science domain plays an important role in all the "big data" research areas. Though many chemists running individual laboratories have not yet seen the direct value of sharing their data, we are beginning to see movement from academia, cooperatives, and publishers.

As data management plans became mandated through funding agencies, researchers turned to research-supporting staff for quick solutions to address the requirements. The research-supporting staff to which the researchers turned consisted of grant officers, institutional repository (IR) service providers, information scientists, graduate students in research groups, and, of course, librarians. Although our librarian job titles and position descriptions did not suggest data management responsibilities, we participated in finding solutions for many reasons: part of our mission is to support institutional researchers; our existing relationships with institutional researchers provide us with an awareness of researchers' data management needs and wants within areas of disciplinary expertise; and our membership in a profession with a tradition of collecting, managing, storing, and making accessible other types of research output affords us valuable insights.

Therefore, we found motivation to engage with campus researchers and began to help with the data storage and sharing needs of researchers at our institution. Importantly, we had administrative encouragement to pursue conversations and discover channels to gain knowledge. With this license, we developed strategies taking advantage of librarians' expertise, campus expertise, and expertise around the world. Now, as our institution is still building our own infrastructure for e-Science, we offer an account of our learning journey including obstacles encountered, resources, and expertise we drew upon, and our approaches to understanding the research community we serve. Our perspective may also help staff from the other research-supporting groups who are just beginning this process to prepare themselves to help meet data sharing needs.

Identify Learning Tools

Once we understand the fundamentals of e-Science and cyberinfrastructure principles, we are ready to learn how to match our own expertise with what researchers need in the disciplinary domains that we support. Since data science is a rather new research domain, foundational literature is not as abundant as in other domains. Relevant research articles are regularly published and should be read, but many other learning channels and learning strategies are available now.

Organizations and Their Publications

Since the emergence of e-Science and cyberinfrastructure, many organizations have dedicated study to data-intensive research issues. Some provide repositories, software, frameworks or other data related tools; some are specially funded organizations dedicated to the evolution of new tools, frameworks and services. Regardless of organizational scope, publications from these organizations are often explicitly shared and can be used as direct learning tools. Earlier we mentioned reports from government agencies and other related organizations for overviews and strategic planning. Here, we list some examples of dedicated research organizations.

- Digital Curation Center (DCC, http://www.dcc.ac.uk/) As the leading center of the United Kingdom's effort on research data management strategy and practice development, DCC hosts a rich open access collection of project documentation, standards, case studies, tutorials and other training documents (7). The topics cover issues most crucial to data sharing and data management. Although the amount of information here may be overwhelming for beginners, we still recommend it as the top resource to consult early in the process.
- Inter-University Consortium for Political and Social Research (ICPSR, http://www.icpsr.umich.edu) With 50 years of experience working with social science data, ICPSR is a leader in data management within and beyond the social science domain. ICPSR is concerned with whole lifecycle data curation, analysis, and access. The consortium sponsors data science research and instruction related to data and maintains a robust data repository. The Digital Curation section (8) and the Guidelines for Effective Data Management Plans section (9) on the ICPSR website are particularly helpful with practical considerations in data management. While the materials are prepared to support Social Science research, research-supporting staff in other research domains will also gain a systematic view of issues to be considered for data management and sharing.
- DataOne (http://www.dataone.org/) and Data Conservancy (http://dataconservancy.org/) The two projects were funded by NSF in 2009 for different focuses – DataONE is devoted to building a framework, cyberinfrastructure, and data repository for environmental science and related fields; the Data Conservancy develops software for data repositories, explores data sharing practices, and fosters development of community, tools, and services for data re-use across social science and science disciplines. Publications on DataOne (10) and Data Conservancy (11) provide good reference articles for future projects; so do the publications listed on the Dataverse Network Project (12). With these publications, individuals can find papers documenting details involved

in the process of making data reusable and translate those details into the domains they support. These resources and tools provided on these sites may also be shared with institutions and researchers.
- Association of Research Libraries (ARL, http://www.arl.org/) For the librarian community, the Association of Research Libraries is concerned with policy and scholarly communication issues that impact libraries. Because e-Science will influence the way scholars communicate and because policy decisions direct the development of cyberinfrastructure, ARL is exploring what librarians can do for e-Science. For example, the ARL study of member institution activity with e-Science and data services provides a sketch of how libraries started services in this area (*13*). What individual subject specialists and other librarians can do in practice is also explored by the members of the ARL e-Science task force (*14*).

Conferences and Workshops

Data science conferences and workshops gather people wishing to communicate recent work, learn from each other, get inspired, and generate and apply new ideas related to data research. Direct communication with peers supporting research data sharing is an efficient way to learn concepts and the research interests of fellow attendees.

The International Digital Curation Conference hosted by DCC (http://www.dcc.ac.uk/events/international-digital-curation-conference-idcc) is one of the largest international events for data science and practice. Whatever your focus or niche, you may find peers working on similar topics at this conference. Presentations and videos from the events are also available on DCC website.

If data sharing is important to a domain, then sessions dedicated to the topic wil likely be found at major conferences for that domain. Librarians may also find valuable programming within the information divisions of domain-specific conferences and within the domain divisions of library association conferences. For example, relevant panels and oral presentation sessions have been organized by Chemical Information Division (CINF) of the American Chemical Society (ACS) at the ACS National Meetings, and by the Division of Chemistry (DCHE) of the Special Library Associations (SLA) at the SLA Annual Conference and Info-Expo for the past three years.

Various institutions and organizations have realized the importance of "train-the-trainers" events and offer workshops to share their expertise in data sharing.

One example is the Data Curation Profile (DCP) Toolkit workshop provided by D. Scott Brandt and Jacob R. Carlson from the Purdue University Libraries (*15*). The workshop series is funded by the Institute of Museum and Library Services (IMLS) to train librarians to interview discipline-appropriate researchers and populate a DCP repository. The accumulated DCPs can be used to reveal the data curation needs of different research communities. Attendees learn how to complete a DCP while thinking through the whole data lifecycle and associated curation issues. Although not all attendees will have the opportunity to execute extensive interviews with researchers, the tool kit provides a framework to organize data-sharing conversations with researchers and successful interviews may help to demonstrate the research-supporting commitment of the librarians conducting the interviews.

ICPSR also provides train-the-trainer style workshops. The ICPSR Summer Program has provided data processing and management training for social scientists around the world since 1963 (*16*). Recently, ICPSR expanded training opportunities with an "Applied Data Science: Managing Research Data for Re-Use" workshop, designed to provide a platform for sharing the expertise from ICPSR, University of Michigan, and all around the world. The workshop combines a big picture overview, detailed case studies, a resource summary, and research updates for the data science field. The balance between discussion of practical issues and up-to-date exploration is the strength of this workshop.

Organizations such as DCC and ICPSR are starting to provide online training programs. While some participants may miss the face-to-face experience that comes with physically attended programs, online training programs offer convenience. Hybrid programs are also emerging such as Data Intelligence 4 Librarians. Developed by 3TU.Datacentrum at Netherlands (http://dataintelligence.3tu.nl/en/home/), the education course provides a mix of online and group meeting learning opportunities. We expect that similar programs will emerge in the U.S. in the near future.

Personal and Organizational Communications

Formal and informal communication with peers can be useful for exchanging problems and stories. When shared, the knowledge and experience of peers can orient research-supporting staff to relevant data sharing issues. Practices that addressed earlier problems may be applied across disciplines and institutions to resolve current problems. Peers are often the best sources of information for surveys and reports previously conducted within our institution. Often, consultation with colleagues is the best way to discover a local institution's history of data management exploration. For instance, a campus wide survey regarding researchers' data management practices conducted by the School of Information at the University of Michigan in 2010 (*17*) was serendipitously discovered by librarians in July 2012 while attending a presentation delivered by the survey author. Compiling such hard-to-find works promotes additional discovery.

RSS feeds and listservs often point to helpful resources and tools. Listservs also provide convenient forums for questions and discussions with peers. Listservs of particular benefit for chemical information specialists include the CHMINF-L (*18*) and SLA-DCHE (*19*). The listserv of Office from the Research Cyberinfrastructure at the University of Michigan (*20*) is crucial for keeping up with activities within our local institutional organization.

Searching online content sharing and social network platforms can also lead to valuable learning channels. Videos, tutorials, and project presentations about e-Science and data sharing are available in abundance via Youtube and Slide Share websites. Following tweets of specific events on data topics and by distinguished researchers in data science on Twitter helps users to stay current. Maintaining a refined list of RSS feeds from interested organizations is another way to keep up to date. For example, RSS feed from the NSF Office of Cyberinfrastrcture Discoveries (http://www.nsf.gov/rss/rss_www_discoveries_oci.xml) provides current information about cyberinfrastructure-enabled discovery.

The tools and channels we mentioned above are those we have used. Beginners are encouraged to explore whichever resources best fit their needs. Resources created for distinct research communities hold significant value for unintended research domains and should not be ignored. In this early era of e-Science, inspiration and critical information may be found in any relevant research and discussions, regardless of the research community for which it was originally intended.

Enrich Your Toolbox

The learning tools and resources described above may be used directly by researchers to manage and inform their data sharing practices and by research-supporting staff to communicate with researchers. Since any of those resources could be more helpful in some contexts over others, it is useful to collect and organize these resources and tools into a toolbox. Table 1 provides a summary of necessary tools and resources. Depending on the needs of the research community, one can consult different resources and tools to address an immediate need while furthering the personal learning process.

Table 1. Tools and resources to be collected to support data sharing

Tools and Resources	Examples
Institutional policy about research data	Research policy pages of various institution(s)
Data management plan (DMP) templates	DMP Online (21), DMP Tool (22), DMP templates by various universities
List of disciplinary repositories	Databib (23), OAD: Data repositories (24), Data Cite: Repositories (25)
Profiling/communication tools	DCP Toolkit (26)
Institutional repositories	DSpace@MIT (27), DataStaR (28), PURR (29)
Metadata standards available	DDI Metadata resources (30), Science Data Literacy Project: Metadata Standards (31), Metadata standards and related resources on D2I Wiki (32), CML (33)
Data citation	DataCite (34), ICPSR: Data Citation (35)
Teaching materials for data literacy	Science Data Literacy Project (36), e-Science Portal for New England Librarians: Science Data Literacy (37)

Besides the tools and resources listed in Table 1, any well-written documents that describe best research/data practices for a relevant research community should be collected and offered up as recommended reading for researchers. Research-supporting staff has opportunities to help research communities improve and perfect their research practices.

Sometimes, the set of tools and resources will not be ideal for certain research community needs. Research-supporting staff must work with researchers and existing tools to develop new, custom tools to fit researchers' needs and disciplinary data re-use needs.

Focus on an Individual Research Community

To identify and prioritize the research-supporting services most essential to a research community, to communicate local research community priorities to institutional stakeholders and leaders, and to make meaningful contributions to infrastructure development, we need to truly analyze the domains and research communities that we serve. Here, we use the domain of Chemistry and the research community connected with the Department of Chemistry at the University of Michigan as examples to demonstrate our approaches.

As mentioned in the introduction to this chapter, some characteristics of the chemistry domain pose unique data sharing problems. The "small science" nature of the domain emphasizes research conducted individually or in small groups. This characteristic may limit researchers' perceptions of the utility and potential of large-scale data sharing within the domain. Although many sub-disciplines of Chemistry focus on individual lab works and traditional publications, an exception exists in the area of crystallography where crystal structures are often deposited to the Cambridge Structural Database (CSD) (*38*) and are curated by staff at the Cambridge Crystallographic Data Center (CCDC). Although direct mining of the crystal structure data can still be challenging due to missing metadata, crystallography is farther along in the data-sharing universe than most other areas of Chemistry.

The relatively proprietary nature of chemistry research poses another barrier to data sharing, since those with ownership stakes in chemistry research stand to lose even as society gains from large-scale data sharing. However, unsustainable increases in research and development cost have prompted increased collaboration between pharmaceutical companies and academic researchers in all stages of their drug development programs (*39*). These collaborations may loosen access to the long-locked gates to some internal databases of those pharmaceutical companies, expedite new drug developments, and reduce R&D cost for pharmaceutical companies eventually. The benefits of large-scale data sharing may then be recognized by stakeholders of this industry.

Sources of Data, Data Types, and Research Profiles in Chemistry

Chemistry data can be found in a variety of sources. Major providers of small molecule data sources are summarized in Table 2. Most Chemistry data generated from research in academia are presented in publications such as journal articles and their supplemental materials, which are then indexed and made searchable in databases and reference books. These databases are relied upon frequently by chemists. Unfortunately, much of the data published in the literature are presented in formats that don't allow for re-use and are provided without associated metadata. Data re-use here is also limited by the high cost of access to the proprietary databases which make data discoverable.

Table 2. Major types of sources for data of small molecules in chemistry

Source	Examples	Metadata	Re-use
Publications	Journal articles and supplemental materials	Buried in texts and captions	No
Proprietary databases indexing data	CAS databases, Reaxys, ASM Phase Diagrams	Limited	Possible but locked up
Reference books indexing data	CRC Handbook of Chemistry and Physics, Springer Materials	Limited	Mostly not
Drug screening databases	ZINC, Internal databases in pharmaceutical companies	Some	Possible except proprietary ones
Open access "hybrid" databases *	ChemSpider, PubChem	Some	Possible
Disciplinary repository	Cambridge Structural Database	Yes	Possible
Institutional repository	DatastaR, PURR	Some, not specific for Chemistry data	Possible

* Content comes from both data mining and users depositing

Chemical Abstract Services (CAS) databases (*40*) and Beilstein database (now a part of Reaxys (*41*)) have existed (first as print, then as electronic databases) for over a hundred years and cover literature back to 18th century. These databases provide a valuable service by extracting data from static publications and making them discoverable. The original purpose of these databases was to make the data and associated publications discoverable, but not reusable, either by human or machine. Technology may offer new potential for content re-use within these rich collections if the issue of costly access could be resolved. Reference books index data similarly to these databases. Databases like ZINC (*42*), which is designed for drug screening, have the most "big data" re-use potential. Significant access barriers exist for most other drug-screening databases containing experimental data due to proprietary interests of large pharmaceutical companies while free databases like ZINC contain mostly data from theoretical calculation.

The potential for data re-use is strongest within the last three sources listed in Table 2. Open access hybrid databases like ChemSpider (*43*) and PubChem (*44*) are designed with different research communities in mind but both emphasize work with small molecules. These two resources share a common strength in policies allowing the public to deposit data. A large amount of work must be dedicated to the curation of the publicly-deposited data before it becomes truly reusable, so resources must be allocated to that curation. Disciplinary and institutional repositories are still in their early stages of development and use. Repository developers are exploring reasonable preservation and access models, especially with regards to metadata standards, to allow effective re-use of chemistry data in these repositories.

A brief discussion of the data of small molecules does not represent the interdisciplinary nature of current Chemistry research and it does not address the importance of the data of polymers and biological molecules in the domain. We expect that future disciplinary repositories will be based upon research themes, such as Energy Science, more often than upon traditional disciplines like Chemistry, because data types and purposes of re-using data are more homogeneous within the same research theme than those across the traditional disciplines. In addition, researchers working on similar research questions tend to participate in active research communities and will be inclined to deposit data where it will be most useful to those working on answering similar research questions.

To illustrate the heterogeneity we have in traditional disciplines like Chemistry, we present a non-exhaustive representation of sample data types associated with Chemistry in Table 3.

Based on the sample data types presented in Table 3, a repository inclusive of all data types that could accommodate preservation and re-use requirements would pose extraordinary problems related to metadata standardization, accessibility, and interoperability. The challenge to create a uniform metadata standard for all the data types here, especially for the metadata describing provenance and experimental conditions, demonstrate the problems associated with this type of repository.

Universities and other institutions are trending toward grouping their researchers by research themes within or beyond traditional departments. The Chemistry Department at the University of Michigan is presented as one example. Figure 1 depicts how principal investigators (PIs) are distributed in both the traditional disciplines and research themes. The data used to plot Figure 1 are summarized from https://www.chem.lsa.umich.edu/chem/faculty/research.php in July 2012. From the bottom graph (based on research themes) in Figure 1, we can see that the majority of PIs have research focuses related to biochemistry, energy, and imaging. PIs often have interest in multiple research themes, which implies that data collected in his/her lab may be of interest to multiple research communities. When we think about data curation for the data produced by these research themed groups, we must consider the data needs of intersecting research communities.

Table 3. Sample small molecule data types in chemistry

Sub-domain	Example Data Types
Synthetic Chemistry	
Preparation procedure	Text with chemical names and special symbols, Scheme
Substance	Identifier, Structure
Characterization/purification	
Spectroscopy	1D and 2D NMR/IR/Mass Spectra, UV-Vis spectra, Atomic Absorption Spectra, Fluorescence Spectra, Raman Spectra
Numerical Data	Boling point, melting point, solubility, etc.
Chromatography	HPLC, GC, CE
Crystallography	Crystallographic structure, Crystal preparation
Computational Chemistry	Gaussian log files
Microscopy	Photomicrograph SEM Image/Video TEM Image/Video AFM Image/Video Confocal microscopy Image/Video
Electrochemistry	Standard electrode potential, Resistance, Voltammetry, Coulometry
Physical chemistry	
Thermodynamics	Entropy, enthalpy, etc.
Kinetics	Reaction rates etc.
Surface chemistry	Adsorption coefficient, etc.

To understand which data types are generated most frequently within our Chemistry Department, we run an ongoing study to profile data types found within publications authored by PIs in the Department. Publications meeting certain criteria are retrieved from the Web of Science index, grouped by PI, and made into a reference set. A FileMaker database was established to hold information extracted from the reference set, including bibliographical information and occurrences of data types that appeared in figures, tables, texts, captions, and especially supplement materials. A controlled vocabulary is in development to describe the data types in consultation with researchers in the Chemistry Department.

The results of this data profiling study will be reported separately. Here are some preliminary statistics about the reference set we collected: 635 journal articles and edited book sections published between 2010 and May 2012 are authored by PIs in the Department. 51 of these articles represent collaborations between two PIs and four are written among three PIs. Ten out of the twelve journals in which our PIs published most are ACS Publications, which means that the availability of data from these publications is highly dependent on the publishing practice of ACS Publications. If publishers like ACS Publications encourage or require data publishing associated with articles as they have done for crystallography data, researchers will have additional incentive to integrate data publishing as a part of scholarly communication thus fostering data sharing.

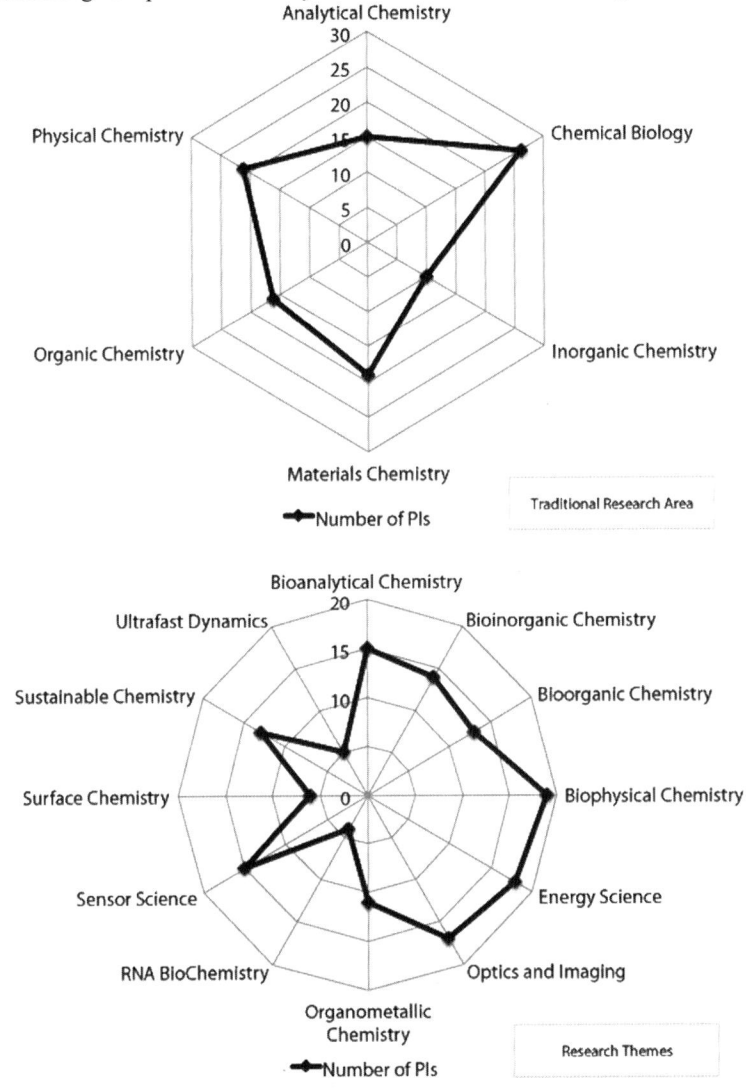

Figure 1. Research Profiles of Chemistry Department at the University of Michigan in Traditional Research Areas and in Research Themes.

In this instance, data types were profiled through publication analysis as an alternative to conducting Data Curation Profile (DCP) interviews with PIs in the Chemistry Department. Our approach reveals the overall departmental profile more directly while DCP provides the complete data story for an individual research projects. Given sufficient time and resources, a combination of the two approaches will provide perspectives at both macro and micro levels.

Metadata

Metadata and metadata standards are crucial for data preservation and sharing. To ensure proper long-term preservation, accessibility and reusability, we need a minimum of descriptive, administrative, and structural metadata (*45*).

Below, we examine a few current disciplinary repositories used for Chemistry data to see how data formats and metadata needs are handled. The results are summarized in Table 4.

As shown in Table 4, three characteristics persist among the examined repositories: (1) data types are limited to crystallography, spectra, structure and some reaction data; (2) data formats are not always suitable for long-term preservation; (3) the amount of metadata required for deposit is minimal and limited to bibliographic and technical metadata. Despite the room for improvement, some encouraging trends are emerging. Two of the repositories request description of reaction conditions and experimental details. These descriptions may be annotated with markup language like XML and become machine-readable descriptive metadata. ChemSpider also asks for explanation of characterization data, which can be annotated into structural metadata to show the relationship among the characterization data and the identifiers. Finally, elective embargo periods are becoming important components of the administrative data for a couple of the repositories. In all, it seems that repositories are lowering barriers to deposit by requesting minimal metadata from depositors. It may be a good strategy to jumpstart repository population, but more systematic collection of metadata and better metadata standards will benefit more data sharing long-term. This strategy is consistent with what Jane Greenberg *et al* described as best practices for a scientific data repository in their 2009 publication (*46*) based on practices of Dryad repository, which was designed for evolutionary biology, ecology, and related disciplines.

In Chemistry, Chemical Markup Language (CML) was an early success for describing data for the semantic web (*33*). Software such as the Microsoft Chem4Word has integrated CML in the package (*47*). Currently, CML supports molecules, compounds, reactions, spectra, crystals and computation chemistry.

Although CML is not designed to be a metadata standard for data repositories, it is an excellent candidate to become a standard for data repositories in Chemistry. In fact, the first two repositories listed in Table 4, eBank-UK eCrystal and SPECTRa have already used subsets of CML to encode metadata and allowed direct export of metadata as CML files.

Table 4. Data format and requested metadata elements of selected repositories/projects in chemistry

Project/ Repository	Data type explored	Standardized data format	Metadata requested when depositing
eBank-UK eCrystal (*48*)	Crystallography	CIF, HKL files	• Bibliography • Data collection parameters • Stages of the structure determination • Experimental conditions
SPECTRa (*49*)	Crystallography	CIF files	• Extracted from CIF, JCAMP, and Gaussian file • Bibliography • Embargo period
	NMR	JCAMP-DX and MDL mol files	
	Computational Chemistry	Gaussian Archive files	
Cambridge Structural Database (*38*)	Crystallography	CIF, FCF or HKL files	• Extracted from CIF • Bibliography • Associated publication • Keywords about study
ChemSpider (*43*)	Chemical structure	MOL, SDF, CDX, SKC files	• General description • Identifiers • Links to websites or publications
	Spectra (¹H NMR, ¹³C NMR, IR and Mass)	JCAMP-DX or –JDX files JPG or PNG for 2D NMR	• Extracted from JCAMP file • Link to associated webpage • Experimental details in Comments field
	Synthetic reaction and associated characterization data	TXT , ChemDraw ChemSketch or RXN file as well as GIF or PNG file for Scheme	• Bibliography • Embargo period • Chemicals involved • Link to publications • Experimental details in Comments and Multimedia fields • Explanation about characterization data • Reaction keywords

Many other markup languages (*50*), such as ThermoML, AnIML, UnitsML have the potential to be used together with CML to create proper metadata standards in Chemistry. We should note that these markup languages are only useful for descriptive and some structural metadata of data in Chemistry. Repositories still need to amend administrative and structural metadata in practice, especially those related to regulation compliance and privacy protection. The creation of a master suite of metadata standards for major data types in Chemistry would benefit localized metadata standards. Various repositories built around different research themes could adopt subsets of the master standards to use in combination with locally developed metadata requirements.

What Do Researchers Really Need?

Regardless of research community, the following questions help research-supporting staff understand the data sharing needs of their constituents. What is the research profile of the department? What types of data are important? What metadata are necessary for various research themes and data types in the department? How can we identify potential data consumers, the designated community, for data from this community? Are there existing repositories to be recommended to our researchers? Will depositing into an institutional repository support this community? How do research practices in the department influence data sharing? Can research-supporting staff help to improve the research workflow so that data sharing becomes effortless and effective? Answers to these questions may take years to find. One urgent question, however, begs to be answered right now: what do researchers really need? This question can be addressed in two means: one practical and one ideal.

The practical approach examines the whole research lifecycle from idea formulation to proposal writing, project planning, data collection, data processing, publishing and sharing, and back to idea formulation to see which parts of the process have been supported by institutional facility and personnel. Meanwhile, we need to consider how the data lifecycle, the data curation cycle, and the scholarly communication cycle can be integrated with the research lifecycle. If any areas not currently being served are identified in the cycles or the integration process of the cycles, these gaps are where future services should be focusing on. This approach has been described by Jacob Carlson from Purdue University at an ICPSR workshop in July 2012. One advantage of this model is that it can be used for institutional strategic planning and also can be used by research-supporting staff to prioritize tasks to support research communities. The model also allows for improvements to the quality of research-supporting services as a whole instead of narrowly focusing on data related tasks.

Based on our own research experience and communication with researchers, the ideal world would involve highly automated workflows enabled by sophisticated lab management systems. Using such systems, any meaningful activities in the lab, from idea being generated, to experiment process, data processing, and paper writing, would be facilitated, recorded and curated with semantic annotations. Then, the content can be selectively and directly shared with designated communities. Backup and preservation of all content would happen behind the scenes without extra effort made by researchers. Since the system and the workflows would be standardized and interoperable, anyone with a need to re-use the data could precisely extract the shared data. Everything shared would be shared with context and recorded provenance creating an ideal environment for data re-use. Labs around the world would essentially be one lab with different rules for different components. The ideal world is far away but possibilities are already emerging. In the domain of Chemistry, a series of projects, including CombeChem, Smart Tea, R4L, e-Bank and e-Crystals, are led by a group of UK scientists (*51*) to create lab management systems similar to what we described. Components of these systems are in development under the umbrella of the Smart Research Framework (SRF) collaborative systems (http://www.mylabnotebook.ac.uk/). Technology revolutions may enable us to realize this ideal research world sooner than we can imagine. Assisting researchers to cultivate good lab practices with the data-centered paradigm in mind will prepare them for the exciting new era in Science.

Learn, Teach, and Collaborate Simultaneously

As a part of teaching and research supporting system of universities, librarians are simultaneously learning and teaching new knowledge and skills as well as collaborating with faculty, students, and staff across campus to accomplish various projects. The emergence of e-Science and data sharing is an opportunity for us to provide new services through the same means. We can apply our expertise in organizing, archiving, and preserving information as well as our traditional roles as connectors of different disciplines on campus. We are nurturing our new expertise in supporting the research cycle, data lifecycle, scholarly communication cycle, and curation of all scholarly processes and outputs. We hope our shared experiences here orient and inspire beginners to get started with this exciting exploration.

Acknowledgments

We would like to thank the University Libraries at the University of Michigan for all of the learning opportunities provided to us. Thanks also to our colleagues, faculty, students, and staff at the University of Michigan for their support for our work.

References

1. Atkins, D. E.; Droegemeier, K. K.; Feldman, S. I.; Garcia-Molina, H.; Klein, M. L.; Messerschmitt, D. G.; Messina, P.; Ostriker, J. P.; Wright, M. H. *Revolutionizing Science and Engineering through Cyberinfrastructure: Report of the National Science Foundation Blue-Ribbon Advisory Panel on Cyberinfrastructure*; Office of Cyberinfrastructure National Science Foundation, January 2003.
2. Glod, A. Cyberinfrastructure, Data, and Libraries. Part 1: A Cyberinfrastructure Primer for Librarians. *D-Lib Magazine*, 2007. http://www.dlib.org/dlib/september07/gold/09gold-pt1.html (accessed July 2009).
3. Gold, A. Cyberinfrastructure, Data, and Libraries. Part 2: Libraries and the Data Challenge: Roles and Actions for Libraries. *D-Lib Magazine*, 2007. http://www.dlib.org/dlib/september07/gold/09gold-pt2.html (accessed July 2012).

4. Garritano, J. R.; Carlson, J. R.. A Subject Librarian's Guide to Collaborating on e-Science Projects *Issues in Science and Technology Librarianship*, Spring 2009. http://www.istl.org/09-spring/refereed2.html.
5. Velden, T.; Lagoze, C. Communicating chemistry. *Nat. Chem.* **2009**, *1* (9), 673–678.
6. Hey, A. J. G.; Tansley, S.; Tolle, K. M. *The Fourth Paradigm: Data-Intensive Scientific Discovery*; Microsoft Research: Redmond, WA, 2009.
7. Resources for Digital Curators. http://www.dcc.ac.uk/resources (accessed July 2012).
8. ICPSR: Digital Curation. http://www.icpsr.umich.edu/icpsrweb/content/ICPSR/dmp/index.html (accessed July 2012).
9. ICPSR: Guidelines for Effective Data Management Plans. http://www.icpsr.umich.edu/icpsrweb/content/ICPSR/dmp/index.html (accessed July 2012).
10. DataONE Publications. http://www.dataone.org/publications (accessed July 2012).
11. Data Conservacy: Browse Our Collection of White Papers and Publications. http://dataconservancy.org/library/ (accessed July 2012).
12. Dataverse Network Project: Publications. http://thedata.org/publications (accessed July 2012).
13. Soehner, C.; Steeves, C.; Ward, J. *E-Science and Data Support Services: A Study of ARL Member Institutions*; Association of Research Libraries: Washington, DC, 2010.
14. Gabridge, T. The last mile: Liaison roles in curating science and engineering research data. *Research Library Issues: A Bimonthly Report from ARL, CNI, and SPARC* **2009** (265), 15–21.
15. DCP Toolkit Workshops. http://datacurationprofiles.org/workshops_content (accessed July 2012).
16. Summer Program in Quantitative Methods of Social Research. http://www.icpsr.umich.edu/icpsrweb/sumprog/index.jsp (accessed July 2012).
17. Fear, K. "You Made It, You Take Care of It", Data Management as Personal Information Management. *Int. J. Digital Curation* **2011**, *6* (2), 53–77.
18. Chemical Information Sources Discussion List. http://www.lsoft.com/scripts/wl.exe?SL1=CHMINF-L&H=LISTSERV.INDIANA.EDU (accessed July 2012).
19. SLA Chemistry Division: DCHE Listserv. http://chemistry.sla.org/listserv/ (accessed July 2012).
20. University of Michigan Office of Research Cyberinfrastructure: News + Events. http://orci.research.umich.edu/news-events/ (accessed July 2012).
21. DMP Online. https://dmponline.dcc.ac.uk/ (accessed July 2012).
22. DMP Tool. https://dmp.cdlib.org/ (accessed July 2012).
23. Databib. http://databib.org/index.php (accessed July 2012).
24. Open Access Directory (OAD): Data Repositories. http://oad.simmons.edu/oadwiki/Data_repositories (accessed July 2012).
25. DataCite: Repositories. http://www.datacite.org/repolist (accessed July 2012).

26. Data Curation Profile Toolkit. http://datacurationprofiles.org/ (accessed July 2012).
27. DSpace@MIT. http://dspace.mit.edu/ (accessed July 2012).
28. DataStaR. http://datastar.mannlib.cornell.edu/ (accessed July 2012).
29. Purdue University Research Repository (PURR). https://research.hub.purdue.edu/ (accessed July 2012).
30. DDI Medtadata Resources. http://www.ddialliance.org/metadata-resources (accessed July 2012).
31. The Science Data Literacy Project: Metadata Standards. http://sdl.syr.edu/?page_id=32 (accessed July 2012).
32. Data to Insight Center: Matadata Standards and Related Materials. http://d2i.indiana.edu/wiki/Metadata_Standards_and_Related_Materials (accessed July 2012).
33. Chemical Markup Language. http://www.xml-cml.org/ (accessed July 2012).
34. DataCite. http://www.datacite.org/ (accessed July 2012).
35. ICPSR: Data Citation http://www.icpsr.umich.edu/icpsrweb/ICPSR/curation/citations.jsp (accessed July 2012).
36. The Science Data Literacy Porject : Educator Resources. http://sdl.syr.edu/?page_id=15 (accessed July 2012).
37. e-Science Portal for New England Librarians: Science Data Literacy. http://esciencelibrary.umassmed.edu/sci_data_literacy (accessed July 2012).
38. Cambridge Structural Database (CSD). http://www.ccdc.cam.ac.uk/products/csd/ (accessed July 2012).
39. Coles, L. D.; Cloyd, J. C. The role of academic institutions in the development of drugs for rare and neglected diseases. *Clin. Pharmacol. Ther.* **2012**, *92* (2), 193–202.
40. Chemical Abstract Services (CAS) Databases. http://cas.org/expertise/cascontent/index.html (accessed July 2012).
41. Reaxys. https://http://www.reaxys.com/info/ (accessed July 2012).
42. Irwin, J. J.; Sterling, T.; Mysinger, M. M.; Bolstad, E. S.; Coleman, R. G. ZINC: A free tool to discover chemistry for biology. *J. Chem. Inf. Model.* **2012**, *52* (7), 1757–1768.
43. ChemSpider. http://www.chemspider.com/ (accessed July 2012).
44. PubChem. http://pubchem.ncbi.nlm.nih.gov/ (accessed July 2012).
45. Green, A.; Macdonald, S.; Rice, R. *Policy-Making for Research Data in Repositories: A Guide*; Data Inormation Specialists Committee-UK: U.K., 2009.
46. Greenberg, J.; White, H. C.; Carrier, S.; Scherle, R. A metadata best practice for a scientific data repository. *J. Libr. Metadata* **2009**, *9* (3−4), 194–212.
47. CML-Aware Software. http://www.xml-cml.org/tools/software.html (accessed July 2012).
48. Coles, S. J.; Frey, J. G.; Hursthouse, M. B.; Light, M. E.; Milsted, A. J.; Carr, L. A.; DeRoure, D.; Gutteridge, C. J.; Mills, H. R.; Meacham, K. E.; Surridge, M.; Lyon, E.; Heery, R.; Duke, M.; Day, M. An e-science environment for service crystallography: From submission to dissemination. *J. Chem. Inf. Model.* **2006**, *46* (3), 1006–1016.

49. Downing, J.; Murray-Rust, P.; Tonge, A. P.; Morgan, P.; Rzepa, H. S.; Cotterill, F.; Day, N.; Harvey, M. J. SPECTRa: The deposition and validation of primary chemistry research data in digital repositories. *J. Chem. Inf. Model.* **2008**, *48* (8), 1571–1581.
50. Nic, M. Chemical XML Formatting. In *Chemical Information Mining: Facilitating Literature-Based Discovery*; Banville, D. L., Ed.; CRC Press: Boca Raton, FL, 2009; pp 99−119.
51. Frey, J. Curation of laboratory experimental data as part of the overall data lifecycle. *Int. J. Digital Curation* **2008**, *3* (1), 44–62.

Editors' Biographies

Norah Xiao

Norah Xiao has over 12 years of research experience in the field of chemistry and 5 years professional experience in the field of Scientific, Technical, and Medical (STM) information science. Norah Xiao is currently with the American Chemical Society as Manager of Editorial Development for Asia.

Leah Rae McEwen

Leah Rae McEwen is the Director of the Edna McConnell Clark Physical Sciences eLibrary and the Chemistry Librarian at Cornell University. Her interests are in electronic-based libraries and specialized information resources and services supporting chemistry and science and technology studies. She is responsible for the programmatic transition and development of an electronic library model for the physical sciences related research and learning communities across Cornell. She has contributed to and served in advisory capacity for a number of information resources and programs deployed by Cornell University and the American Chemical Society (ACS). She currently serves as the Secretary for the ACS Division of Chemical Information and as a member of the ACS Joint Board-Council Committee on Publications. She holds Masters of Science degrees in Library Science (Emporia State University) and Nutritional Biochemistry (Cornell University).

Indexes

Author Index

Allard, S., 47, 97
Blake, C., 97
Brach, C., 129
Hswe, P., 115
Kafel, D., 69
Li, Y., 145
Martinsen, D., 31

Palmer, C., 97
Pryor, G., 1
Schlembach, M., 129
Stanton, J., 97
Tschirhart, L., 145
Williams, A., 19
Xiao, N., ix

Subject Index

A

ACS. *See* American Chemical Society (ACS)
American Chemical Society (ACS), 34
ANDS. *See* Australian National Data Service (ANDS)
Australia, national data management initiatives, 48
Australian National Data Service (ANDS), 48

C

Chemical Abstracts Service, 25
Chemistry data
  sources, 152, 152*t*
  types, 152, 154*t*
Chemistry research profiles, 152, 155*f*
ChemSpider, 19, 20, 21*f*
  SyntheticPages, 23, 24*f*
  vs. Chemical Abstracts Service, 25
Compound centric community resource
  ChemSpider, 19, 20, 21*f*
  SyntheticPages, 23, 24*f*
  vs. Chemical Abstracts Service, 25
  Google Patents, 23*f*
  Learn Chemistry wiki, 26*f*
  overview, 19
  SpectraSchool website, 27*f*

D

DANS. *See* Data Archiving and Networked Services (DANS)
Data and publication life cycles, 72*f*
Data Archiving and Networked Services (DANS), 50
Databib, Purdue University, 134
Data Conservancy and the Data Observation Network for Earth (DataONE). *See* DataONE
Data curation
  business aspirations, 5
  citations, 12
  competition, 4
  data dependency, 9
  grant providers, 3
    EPSRC research grant value, 4*f*
  impact, 7
  motives, 14
    complexity, 14
    management, 15
    policy, 14
    scale, 14
  openness, 7
  overview, 1
  research governance enhancement, 11
  research incentives, 5
  techniques, 9
  traditions, 9
Data Curation Profiles, Purdue University, 134
Data format, individual research community, 157*t*
Data management curriculum, 87, 89*f*
Data Management Planning (DMP) Tool, 140
Data Management Plan Self-Assessment Tool, Purdue University, 134
Data management services, libraries. *See* library data management services
DataONE
  challenges, 60
  collaborations, 63
  community engagement activities, 62
  context, 54
  cyberinfrastructure, 59
  data lifecycle, 58, 59*f*
  mission, 56
  organizational structure, 56
  overview, 47
  stakeholders, 62
  synergies, 63
  vision, 56
  working groups, 58
DCC. *See* Digital Curation Centre (DCC)
Deutsche Forschungsgemeinschaft (DFG), 49
DFG. *See* Deutsche Forschungsgemeinschaft (DFG)
Digital Curation Centre (DCC), 51
Discipline repositories, research data management, 131
DMP Tool. *See* Data Management Planning (DMP) Tool

E

Engineering and Physical Sciences Research Council (EPSRC), 3
EPSRC. *See* Engineering and Physical Sciences Research Council (EPSRC)
EPSRC research grant value, 4*f*
E-science continuing education programs, librarians
 libraries, 71
 New England, 74
  collaborative data repository, 89
  data and publication life cycles, 72*f*
  data management curriculum, 87, 89*f*
  e-science librarianship journal, 90
  IMLS National Leadership Planning Grant, 86
  planning, 81
  portal, 83
  professional development day workshops, 79
  research, 91
  scholarship, 91
  Science Boot Camp, 80, 81*t*
  strategy, library programs, 75, 76*t*
  symposium, 77
  overview, 69
E-science librarianship journal, 90
E-science portal, New England librarians, 83, 140

G

Germany, national data management initiatives, 49
Google Patents, 23*f*

I

IMLS National Leadership Planning Grant, 86
Institutional repositories, research data management, 131
Interdisciplinary data science education
 applications, 100
 continuing education model, 108
 ingredients, 102
 overview, 97
 workshop, 104
  group discussion, 105
  group result synthesis, 106

J

Johns Hopkins University, research data management, 132

L

Learn Chemistry wiki, 26*f*
Library data management services
 librarians' role, 117
 necessity, 117
 overview, 115
 process, 118

M

Massachusetts Institute of Technology, research data management, 133
Metadata, individual research community, 156, 157*t*

N

National data management initiatives
 Australia, 48
 Germany, 49
 Netherlands, 50
 overview, 47
 United Kingdom, 51
 United States, 52
 See also DataONE
National Federation of Advanced Information Services (NFAIS), 32
National Information Standards Organization (NISO), 32
NFAIS. *See* National Federation of Advanced Information Services (NFAIS)
NISO. *See* National Information Standards Organization (NISO)
NISO/NFAIS Working Group, 38
 archiving, 41
 categories, 38
 discoverability, 40
 dissemination, 42
 peer review, 40
 preservation, 41

O

Organizational communications, 149

P

Personal communications, 149
Professional development day workshops, New England librarians, 79
Purdue University, research data management, 134
    Data Curation Profiles, 134
    Data Management Plan Self-Assessment Tool, 134
    Databib, 134
Purdue University Research Repository, 135

R

Research data management
    challenges, 141
    discipline repositories, 131
    DMP Tool, 140
    e-science portal, New England librarians, 140
    institutional repositories, 131
    Johns Hopkins University, 132
    Massachusetts Institute of Technology, 133
    overview, 129
    Purdue University, 134
        Data Curation Profiles, 134
        Data Management Plan Self-Assessment Tool, 134
        Databib, 134
        Purdue University Research Repository, 135
    resources, 142
    University of Illinois at Urbana-Champaign, 140
    University of North Carolina, 136
    University of Virginia, 135
    University Plans, 137t
Research data sharing
    individual research community, 151
        chemistry data, 152, 152t, 154t
        chemistry research profiles, 152, 155f
        data format, 157t
        metadata, 156, 157t
        researchers' requirements, 158
    learning tool identification, 146
    conferences, 148
    organizational communications, 149
    organizations' publications, 147
    personal communications, 149
    workshops, 148
    overview, 145
    tools, 150, 150t
Researchers' requirements, individual research community, 158

S

Scholarship, New England librarians, 91
Science Boot Camp, New England librarians, 80, 81t
SpectraSchool website, 27f
Strategy, e-science library programs, 75, 76t
Supplemental journal article materials
    ACS, 34
    NISO/NFAIS Working Group, 38
        archiving, 41
        categories, 38
        discoverability, 40
        dissemination, 42
        peer review, 40
        preservation, 41
    overview, 31
Symposium, New England librarians, 77

T

The Netherlands, national data management initiatives, 50

U

United Kingdom, national data management initiatives, 51
United States, national data management initiatives, 52
University of Illinois at Urbana-Champaign, research data management, 140
University of North Carolina, research data management, 136
University of Virginia, research data management, 135
University Plans, research data management, 137t